THE HOME
MACHINIST'S
HANDBOOK

THE HOME
MACHINIST'S
HANDBOOK

DOUG BRINEY

TAB Books
Division of McGraw-Hill

New York San Francisco Washington, D.C. Auckland Bogotá
Caracas Lisbon London Madrid Mexico City Milan
Montreal New Delhi San Juan Singapore
Sydney Tokyo Toronto

pbk   32 33 34 35 QFR/QFR 1 5 4 3
hc   2 3 4 5 6 7 8 9 10 QFR/QFR 1 5 4 3 2 1

Library of Congress Cataloging-in-Publication Data

Briney, Doug.
    The home machinist's handbook.

    Includes index.
    1. Workshops—Handbooks, manuals, etc.   I. Title.
TT153.B74   1983   684'.08   83-4951
ISBN 0-8306-0573-8
ISBN 0-8306-1573-3 (pbk.)

Contents

Introduction

A LOT HAS BEEN WRITTEN ABOUT MACHINE tools and how they are used to make parts. Most books and literature on the subject are written in very technical terms for the professional machinist, someone interested in obtaining high-volume production rates at the lowest cost. These books assume that the machinist has unlimited funds and can equip his shop with the very latest and best equipment.

This book is different. It is written for the nonprofessional—the man who wants to work on a few projects at home. This book is written for the hobbyist who wants to make a few parts for his radio-controlled airplane or boat, the model railroader building a new locomotive, or the amateur horologist repairing a clock. It is intended for the inventor who needs a few parts to try out his latest idea, or the engineer who wants to build a prototype for a new product. This book is intended for you—to give you an understanding of the equipment, methods, and materials used in basic machine shop work.

SELECTING TOOLS

A word of caution. Since I don't know your specific needs, I can't tell you what tools you should buy to equip your shop. Your tools will be based on the type of projects on which you intend to work. Naturally, tools that would be required for one type of project may not be required for another. Only you can decide which tools are essential. Since most of you are working with a limited budget, it is necessary for you to select those tools that will give you the best value for your money.

I strongly recommend that as you set up your shop, you take the time to study your needs carefully. Select only the tools that you need. Then, when you buy, avoid cheap bargain tools. Buy tools that will last. If a tool is worth buying, buy a good one.

Years ago, when refinishing a boat, I needed an electric sander. I shopped around and found a bargain—or so I thought. The sander broke down before the boat was half finished. The pot-metal gears in it didn't hold up. I set the sander aside

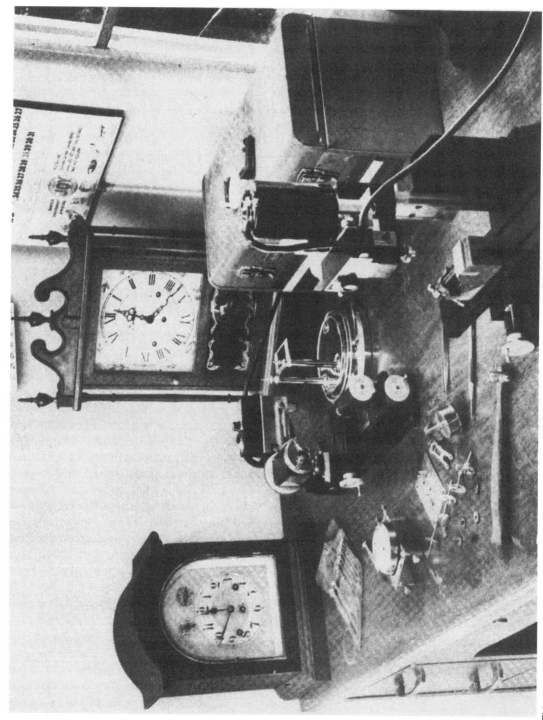

This amateur horologist uses an old office desk as a workbench. A window in front of the desk and an overhead fluorescent light provide plenty of illumination for fine detail work.

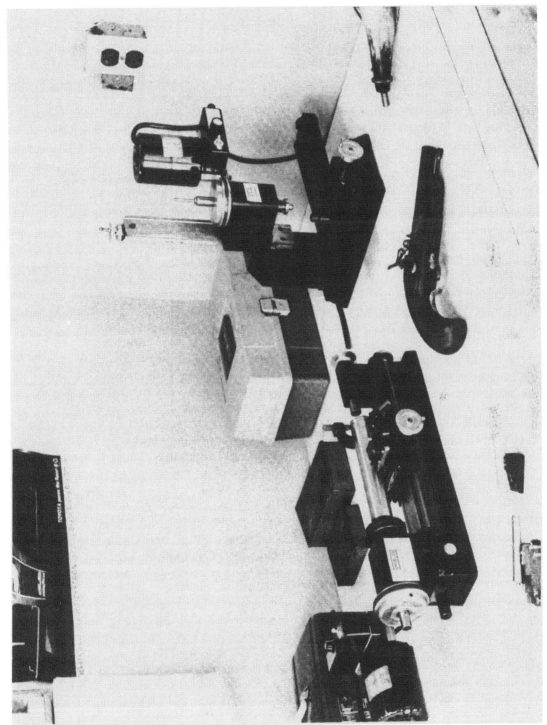

This work area belongs to an amateur gunsmith. The wooden workbench provides plenty of storage space for small parts and a solid platform for machinery. Illumination is provided by a combination of incandescent and fluorescent light fixtures.

thinking that one day I'd replace the gears. Then, to keep my project moving, I went out and bought a new sander. This time, instead of looking for a bargain, I shopped for quality. The new sander finished my boat and all of the jobs that followed. I purchased that sander 23 years ago. I still use it. I never did bother to fix my bargain sander. Eventually it was tossed out. The lesson was a simple one, but one not to be forgotten. When shopping for a tool, look first for quality, then for price.

SAFETY

Some people believe that even the simplest tools cause accidents. That's not true. Tools don't cause accidents; it's the person using the tool that causes the accident. When using any tool, you need to be careful. When using power tools, you must be even more careful. No tool knows the difference between the material you are cutting and you. If you get in the way, you're going to be hurt. Any tool improperly cared for or used can be dangerous. Safety is up to you, not the tool.

I've listed some basic do's and don'ts below, and throughout the book I will point out other precautions that apply to specific tools or machines. Read them carefully, remember them, and most importantly practice them.

● Use your tools properly and keep them in good shape.
● Dress properly. Do not wear loose fitting clothing, ties, necklaces, loose sleeves, etc. Remove your jewelry, rings, wrist watches, etc. Keep your hair short and out of the way.
● Protect your eyes. An eye shield or safety glasses are a must. They should be worn whenever you are hammering, drilling, grinding, or using a lathe or mill. They should also be worn when working with cleaning agents or solvents.
● Know your machine, its capabilities, and its limitations.
● Keep your work area clean and neat.
● Keep your hands away from moving parts.
● Keep your mind on the job. When you're

working, don't talk. When you are talking, don't work.
● Never remove chips with your fingers. Use a brush to clean machinery and equipment.
● Shut the power off before changing setups or making adjustments.
● Don't carry sharp tools in your pockets.
● Keep a first aid kit handy. Always treat and cover cuts.
● Keep the floor clean. Sweep up scraps and chips that might cause you to slip or trip.

SETTING UP A SHOP

The ideal shop has the equipment set up permanently. This is a luxury not everyone can enjoy. Some of you will be limited in space, so that you can only set up the one machine that you are using at that particular time. Whatever your situation, certain basic considerations should be observed. Mount your equipment permanently whenever possible. Establish storage areas for each piece of equipment that is not permanently mounted. Your work area should always be maintained as clean and uncluttered as possible. Your bench should always be free and clear of unnecessary tools.

A clean bench means that any tool or piece of equipment not in use should be returned to its storage area immediately after use. Neatness is a good work habit. It is not only safer to work in an uncluttered area, but you will find that you make fewer mistakes and have fewer problems. You will save time and effort in the long run.

Proper lighting is another important consideration. A good fluorescent light located directly overhead is best. The fluorescent light casts fewer shadows and provides good, bright, nonglare illumination for your work area. Walls, cabinets, bench top, and so on should be painted with a light color semigloss paint. The light color will reflect more light, and the semigloss paint will be easy to keep clean.

The last consideration is cleanliness. Machine tools create a lot of messy chips. A fast turning lathe or mill will scatter an unbelievable amount of chips all over your work area. Open shelves, open containers, open drawers, and so on will collect the chips and make cleanup a lot harder. The best bet is to keep everything covered. Have cupboards instead of open shelves. Arrange your shop so that it will be easy to clean. Avoid inaccessible areas.

Chapter 1

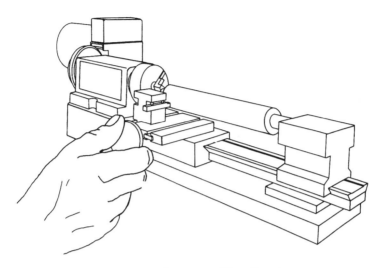

Reading Prints

MACHINED PARTS ALWAYS START OUT AS AN idea in someone's head. This idea, when pictured on a piece of paper, shows the part's shape and defines its size. It is called a *print*. The term print actually refers to a copy of an original drawing, but I will call any drawing that defines a part or an assembly a print. The person who made the drawing or print is the designer. If you are building a part that you designed, you won't have any trouble understanding your print. If you are working to a print that someone else has prepared, however, you will need a good understanding of standard drawing practices.

Normally the designer will indicate on the print all of the information that you will need to successfully produce a part (Fig. 1-1). The print will specify the type of material from which the part is to be made. It will tell you the finished size and will clearly define any special features such as holes, slots, cutouts, and so on. Finally, it will tell you if the part should be heat-treated or if any special finish is required. Items not critical to the performance of the part are sometimes left unspecified,

but generally the designer will put on the print everything you need to know to build the part.

DRAWINGS

If a part is a simple one, the designer may choose to draw it in a three-dimensional drawing. This type of drawing is called a *pictorial drawing*. With this type of drawing it is very easy to visualize the part, but it is difficult to define a complicated part this way (Fig. 1-2). For this reason, most designers use a method of drawing called *orthographic projection*. On a print using orthographic projection (Fig. 1-3) the part is usually shown in three views. A side view, an end view, and a top view are usual. The views are always located on the drawing in relation to each other, so that there can be no question from where the views are taken.

The relationship of views on a drawing can be visualized by imagining the object to be drawn as floating inside a clear plastic box (Fig. 1-4). If you look straight down on the box and trace on its surface the outline of the part inside, you will have a top view. Imagine that you repeat this process for

1

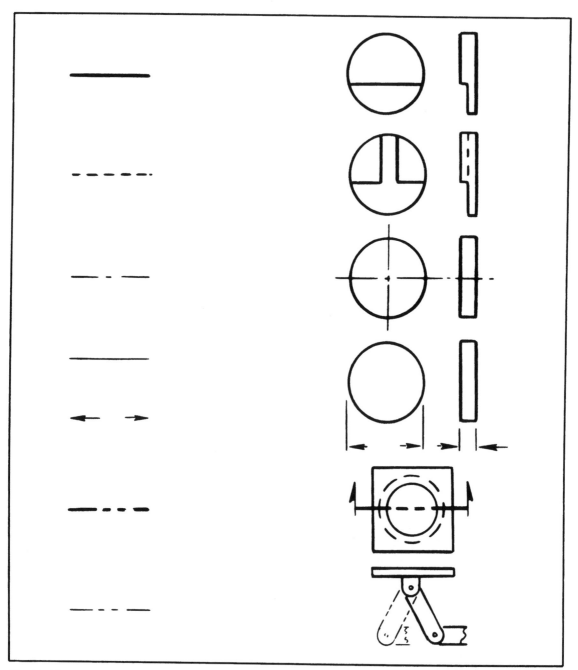

Fig. 1-1. The types of lines used in blueprint drawing have different meanings. A dark solid line, called an object line, is used to show the outline of a part and its features. A dark line broken into a series of short dashes, called a hidden line, is used to show features hidden from view. Center lines indicate the center of round or symmetrical parts or features. Extension lines and dimension lines define the surfaces being measured. Section lines show the cutting plane and direction of viewing when sectional views are required. Phantom lines are used to show alternate positions for movable parts.

2

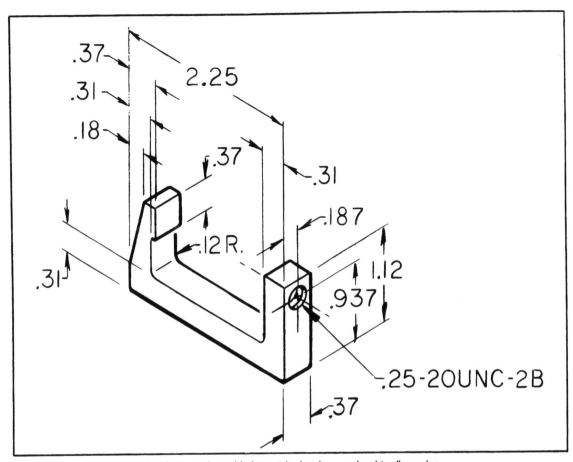

Fig. 1-2. Three-dimensional drawings, such as this isometric drawing, are hard to dimension.

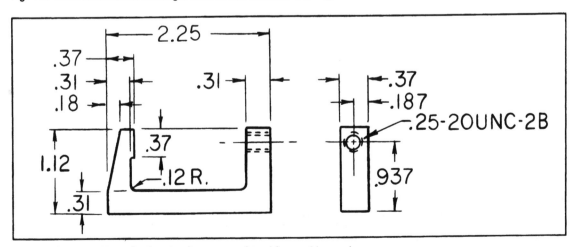

Fig. 1-3. Orthographic projection drawings are preferred for machine work.

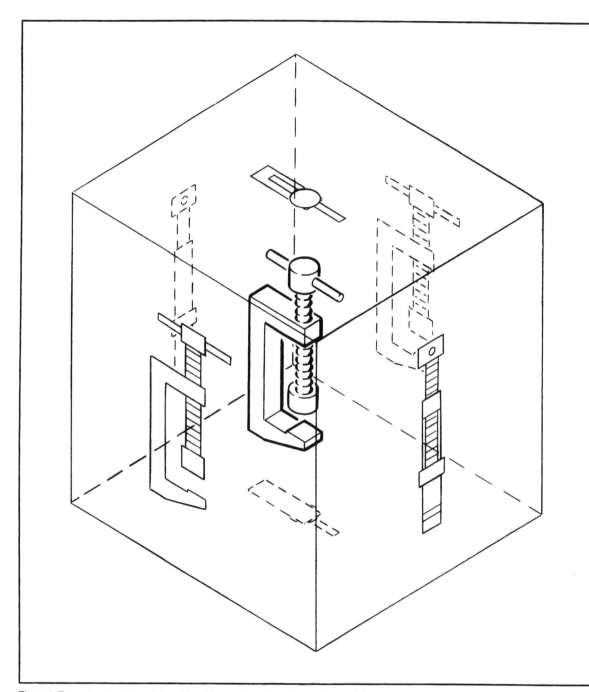

Fig. 1-4. To understand the relationship of the various views in an orthographic projection, visualize the part in a clear plastic box. The outline of the part is projected onto each surface of the box: a top view, four side or elevation views, and a bottom view. If the box was unfolded and laid flat, each of the various views would be located in relation to each other.

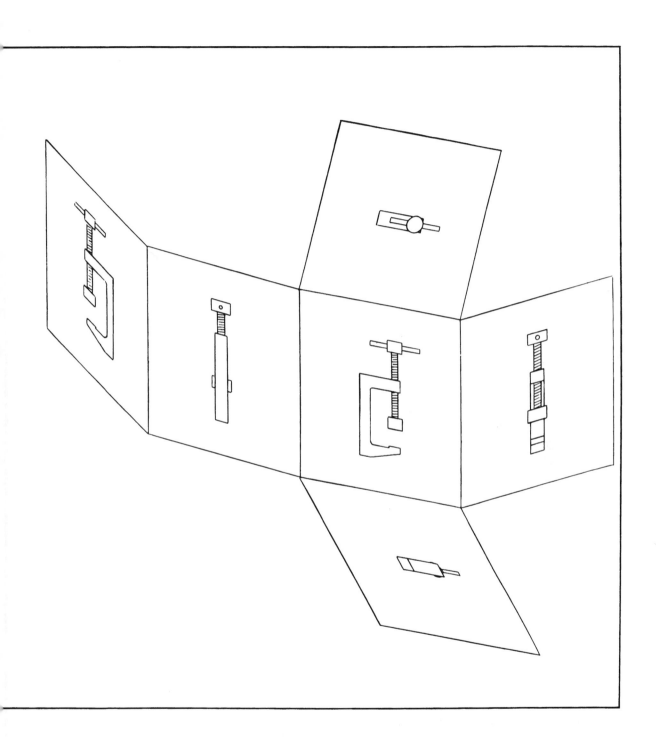

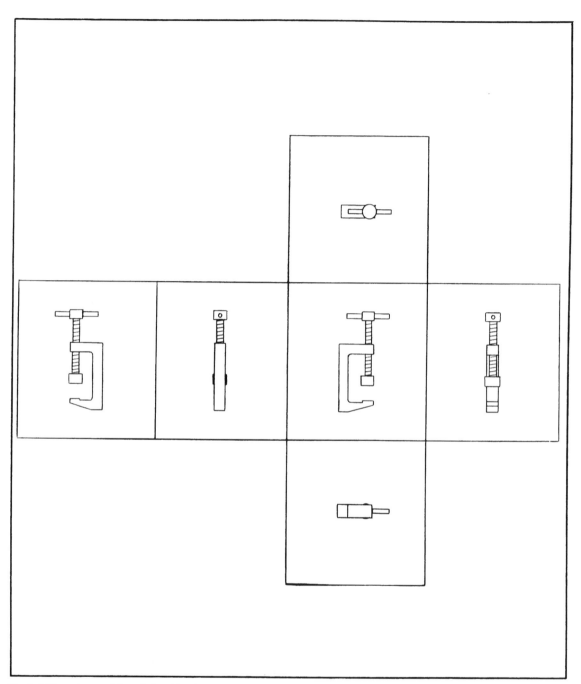

Fig. 1-4. To understand the relationship of the various views in an orthographic projection, visualize the part in a clear plastic box. The outline of the part is projected onto each surface of the box: a top view, four side or elevation views, and a bottom view. If the box was unfolded and laid flat, each of the various views would be located in relation to each other. (Continued from page 5.)

the side view, an end view, and so on. Now, if the sides of the box are hinged so that you can open it up and lay it flat, you will have the various views of the part laid out just as they would be on a print using orthographic projection. Only two views would be shown if the part were a simple one such as a baseplate with a few mounting holes. A third view would be wasted because it wouldn't show anything the first two didn't already show. If the part is a complicated one, more than three views might be shown. The designer will generally show as many views as he needs to clearly define the part.

If the part has a sloping surface with some holes or cutouts in it, the designer will probably show that surface as an auxiliary view. The auxiliary view is projected *normal to* (at 90° to) the surface to be shown, (Fig. 1-5). This will allow the part to be dimensioned exactly as it will be measured on the sloping surface. If the designer had projected the view straight up for a top view, some of the dimensions would have to be determined by mathematical calculation.

Features of a part that are normally hidden from view are shown as broken lines. Figure 1-6 shows a part with a hole drilled through and a counterbore in one end. The hole is located by dimensions to the center lines. The hole diameter, the counterbore diameter, and the depth of the counterbore are called out in a *drilling note*. Dimensions are never shown to a hidden line. If the designer wanted to specify the depth of the counterbore with a direct dimension, he would have shown the part with a section broken away (Fig. 1-7). The diagonal lines are called *section lines* and indicate that a section of the part has been removed to show the inside.

If the hidden feature of the part is sufficiently complicated, the designer may decide to show a *sectional view* (Fig. 1-8). First, the basic orthographic projection views are shown. On one of them, the designer identifies from where he is taking the sectional view. Figure 1-8 shows section A-A taken from the side view.

If the part is cylindrical or symmetrical, the designer may choose to show a quarter section (Fig. 1-9). When this is done, the end view and center lines show that the part is symmetrical. Then, in the side view, the designer simply cuts a fourth of the part away to show the internal features.

Repetitious features, such as threads and gear teeth, take a lot of time to draw. The designer may choose to save time by drawing them in a simplified way using broken lines (Fig. 1-10).

DIMENSIONS

Dimensions usually are not shown directly to the surfaces of the part; they are shown to extension lines instead. Both extension lines and dimension lines are always drawn lighter than the outline of the part. For most machine drawings the dimensions are given in inches, and fractions of an inch are expressed as decimals.

Everyone is familiar with the standard ruler, which divides inches into common fractions such as 1/2, 1/4, 1/8, 1/16, 1/32, and 1/64. This is fine for many applications; however, most machine work is done by dividing the inch in decimal fractions (1/10, 1/100, 1/1000, and 1/10,000). These may be shown as fractions, but it is common practice to show them in decimal form. Decimals are read as follows:

● 1/10 = 0.1 = one tenth of an inch = 0.100 = one hundred thousandths of an inch.

● 1/100 = 0.01 = one hundredth of an inch = 0.010 = ten thousandths.

● 1/1000 = 0.001 = one thousandths of an inch.

● 1/10,000 = 0.0001 one ten thousandths of an inch.

Common fractions can be converted to decimal form by dividing the numerator (upper number) by the denominator (lower number). As an example, you can convert the fraction ⅛ to decimal form by dividing 1 by 8 as follows:

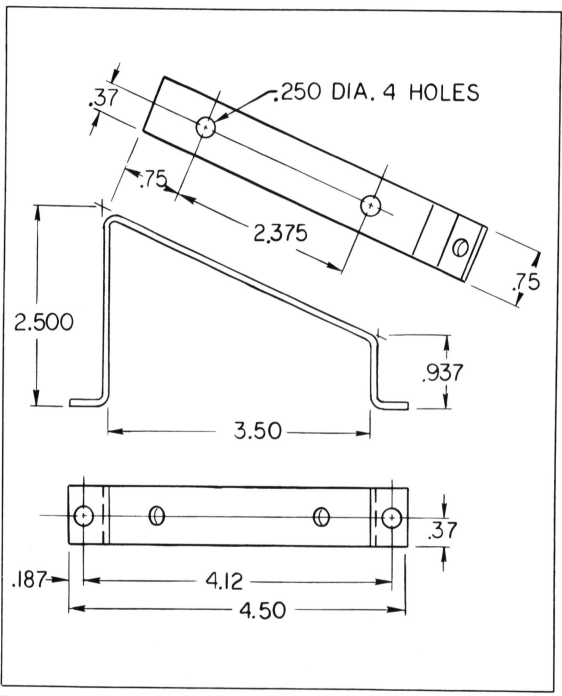

Fig. 1-5. Views of sloping surfaces are usually projected normal to (at 90° to) the sloping surface. These views, called auxiliary views, allow all dimensions shown on the view to be measured directly on the surface of the part.

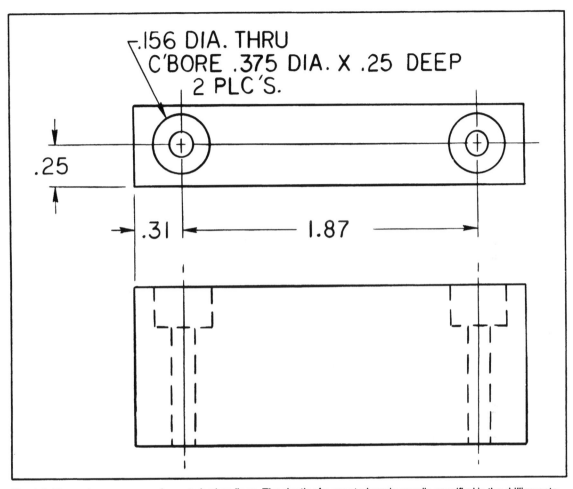

Fig. 1-6. Dimensions are never shown to broken lines. The depth of a counterbore is usually specified in the drilling note.

$$\begin{array}{r} .125 \\ 8\overline{)1.000} \\ \underline{8} \\ 20 \\ \underline{16} \\ 40 \\ \underline{40} \\ 0 \end{array}$$

Thus the decimal equivalent of ⅛ inch is .125 or one hundred-twenty five thousandths. The fraction ¾ would be converted in the same way:

$$\begin{array}{r} .750 \\ 4\overline{)3.000} \\ \underline{28} \\ 20 \\ \underline{20} \end{array}$$

So, the decimal equivalent of ¾ is .750 or seven hundred-fifty thousandths. All common fractions can be converted in the same way so that:

1/2   inch = .500
1/4   inch = .250
1/8   inch = .125

9

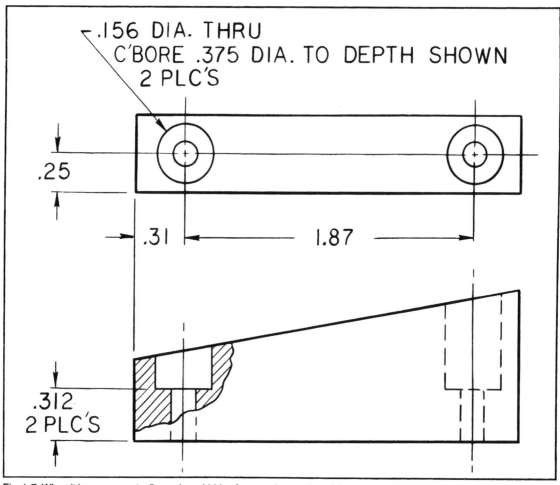

Fig. 1-7. When it is necessary to dimension a hidden feature, that portion of the part hiding the feature may be broken away on the drawing.

1/16   inch = .0625
1/32   inch = .03125
1/64   inch = .015625

A complete fraction to decimal conversion chart is located as a reference in the Appendix. Whole inches are shown to the left of the decimal point, and fractions are shown to the right.

On a print, decimals are generally rounded off to two or three places. As an example, 1/16 of an inch would be shown as 0.06 or 0.063 instead of 0.0625. The fraction 1/32 of an inch would be shown

as 0.03 or 0.031 instead of 0.03125. The number of places that a decimal is carried out indicates the accuracy that the designer wants the machinist to achieve in fabricating the part. This accuracy is normally referred to as the *tolerance*.

TOLERANCES

The tolerance on a dimension tells the machinist how closely to hold the dimension if the part is to work properly. As an example, a dimension might read, 12.50 ± .06. (The sign ± means "plus or minus.")

10

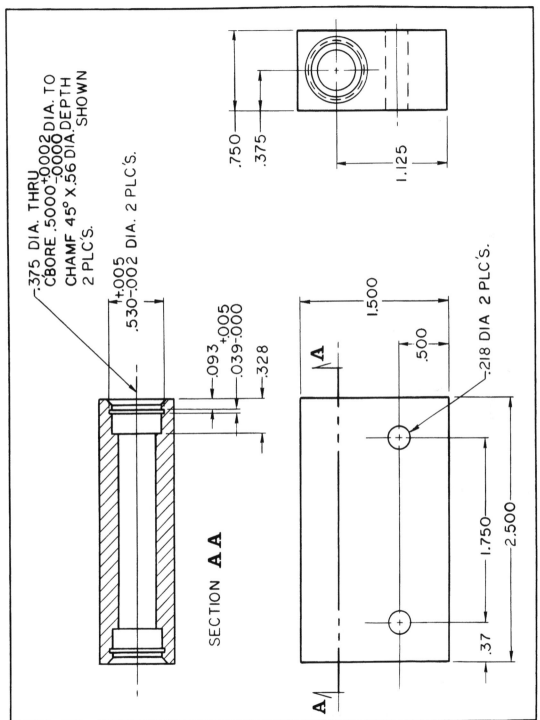

.375 DIA. THRU
CBORE .5000$^{+.0002}_{-.0000}$DIA. TO
CHAMF 45° X .56 DIA.DEPTH
2 PLC'S.   SHOWN

.530$^{+.005}_{-.002}$ DIA. 2 PLC'S.

.093$^{+.005}_{-.000}$

.039$^{+.005}_{-.000}$

.328

SECTION **A A**

.750

.375

1.125

1.500

.500

.218 DIA 2 PLC'S.

1.750

2.500

.37

A

A

Fig. 1-8. If the hidden feature is sufficiently complicated, an entire portion of the part may be cut away on the drawing. The view of the part is then called a sectional view. The section line A-A shows exactly where the part was cut and the direction of the view.

11

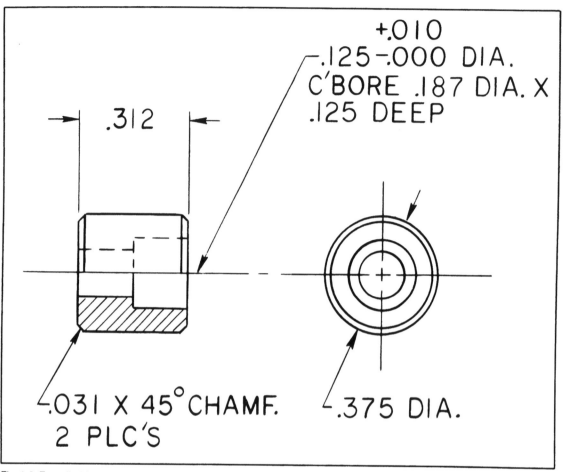

$$+.010$$
.125 −.000 DIA.
C'BORE .187 DIA. X
.125 DEEP

.312

.031 X 45° CHAMF.
2 PLC'S

.375 DIA.

Fig. 1-9. For cylindrical parts with complex internal features, it is not uncommon to show the part with a quarter section removed to reveal the internal features.

Taking the basic dimension of 12.50, you would first add the .06 tolerance and then subtract the .06 tolerance. This would give you two dimensions—12.43 and 12.56. These dimensions are the limits of the tolerance. If you machine the part to any dimension between these limits, it will work satisfactorily. A tolerance of .06 could be used on any part that does not have to mate closely to another part.

If the dimension is locating features that have to match features on a second part, the tolerance would be tighter. Bolt holes used to fasten one part to another might be held ± .010. The tighter the

tolerance, the more difficult it will be for you to make the part. For this reason, the designer will usually specify as large a tolerance as possible. Since it is time consuming for him to write a tolerance next to each dimension on the print, he will usually state the tolerances in a note (Fig. 1-11). This note is sometimes found near the drawing title and sometimes along with other notes on the face of the drawing.

A typical tolerance note would indicate the tolerance based on the number of places in the decimal. As an example the note might read "Tolerances: .XX ± .03, .XXX ± .010, < ± 2."

12

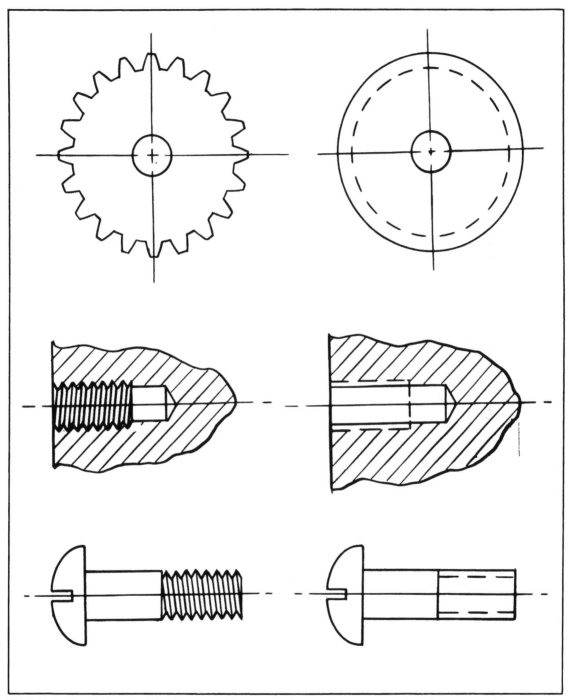

Fig. 1-10. To simplify the task of drawing a part with repetitious features, the presence of these features may simply be shown with a broken line, as illustrated here.

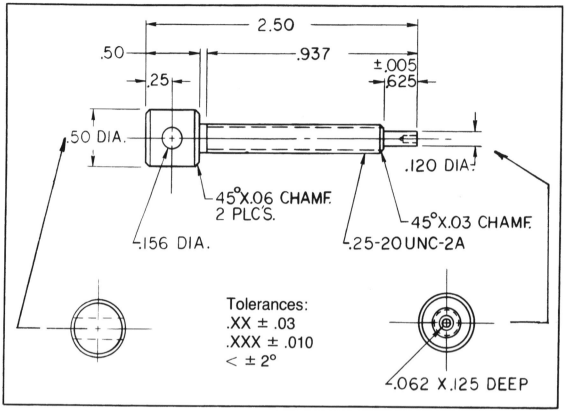

Fig. 1-11. Instead of repeating the tolerance for each dimension on a drawing. Standard tolerances are usually shown in a note. Exceptions to the standard tolerance are the only tolerances shown with a dimension.

This would mean that for any dimension having a two-place decimal, the tolerance is ± .03 inch. For any dimension having a three-place decimal, the tolerance is ± .010 inch. All angles must be held within ±2°. On a drawing with a tolerance note, the only time a tolerance would be shown next to a dimension is when that particular dimension needs to be held closer than normal, such as the diameter of a shaft that has to fit snugly into a bearing. In this case the tolerance might only be a few ten-thousands of an inch. To avoid any confusion, the designer would specify this with the dimension rather than in notes.

Limits

Sometimes a designer may use *limits* (Fig. 1-12) instead of tolerances. When this is done, he has simply added and subtracted the tolerances for you. Instead of dimensioning the diameter of a shaft as .562 ± .005, he would dimension it $\frac{.567}{.557}$. This would indicate to you, the machinist, that any diameter between the two limits. .567 and .557 is acceptable. The larger dimension is called the *upper limit*, and the smaller dimension is called the *lower limit*. Limits are frequently used to dimension critical hole and shaft diameters.

The Importance of Tolerances

Just how important are dimensions and tolerances to you? As a rule, when you are building one-of-a-kind items, the dimensions are only a guide to follow. If you drill a hole a little oversize there's no problem. Just make the part that fits into it a little oversize to match. If the spacing between holes on one part is a little off, you can space the holes in the mating part the same so that they will match. In other words, the dimensions and tolerances can be varied to suit your needs. On the other hand, if you are making a part that has to mate an existing part or a commercial part, it's a different story.

When making a part that has to mate to an already existing part, it is very important to make the dimensions match, particularly if the part is a commercial part such as a bearing. Bearings are manufactured to very close tolerances. You want the bearing to be located properly and to fit snugly so that it won't turn in your part. In this case it's very important for you to hold the designer's tolerances. If you don't, you're going to have problems.

I think that the best practice is to always try to

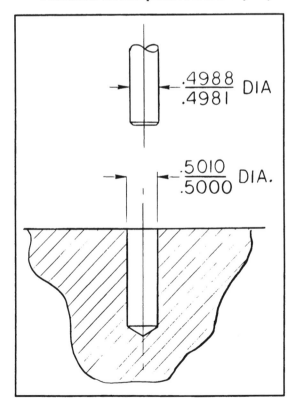

Fig. 1-12. Limit dimensions are sometimes used in place of tolerances.

follow the dimensions and tolerances established by the designer. Then, if you have a problem, you can always modify the design to suit your needs. If you get into the habit of following the print, the job will go much smoother in the long run.

METRICS

This chapter would not be complete without a few words about the metric system of measurement. For linear measurements, the English system of measurement is based on the inch; the metric system is based on the meter. One meter is equal to 39.370 inches. Because this unit is so large, it is convenient to divide it into smaller units. These units are based on divisions of ten:

1 meter = 10 decimeters = 39.370 inches
1 decimeter = 10 centimeters = 3.937 inches
1 centimeter = 10 millimeters = .3937 inches
1 millimeter = .03937 inches

The standard metric unit of measure that is seen most frequently on machine prints is the millimeter. (A metric-to-English conversion chart converting inches to millimeters is located in the Appendix.)

PLANNING

The print usually will not tell you which operations to do first when machining a part. You will have to plane the sequence of machine operations yourself. The best way to do this is to study the part and then mentally visualize each of the machine operations. Decide how you are going to hold the part for each operation. A few moments spent mentally machining a part can save you a lot of time and trouble later on.

Each part has its own special considerations. As a rule of thumb, the normal sequence of operations for a lathe part is as follows:

1. Cut the raw material to length. Allow excess length for chucking, center drilling, etc.
2. Machine internal features, if any. If the part is long, it should have at least one end center-drilled to help support it in the lathe.

15

3. True the outside diameter.

4. If the part is cylindrical, or has no features at 90° to its axis, it may be turned to its finished size and shape.

5. If the part is not cylindrical or has features machined at 90° to its axis, either these features must be machined now, or provisions for locating and holding the part during the required operations must be made now.

6. Finishing operations, such as polishing, should be performed now prior to removing excess chucking material.

7. Part or remove excess chucking material.

Milling operations differ from lathe operations in that the internal features are generally machined after the external features are machined. This allows the part to be clamped on machined surfaces. Drilling and tapping holes are usually the last machine operations done on a milled part.

Chapter 2

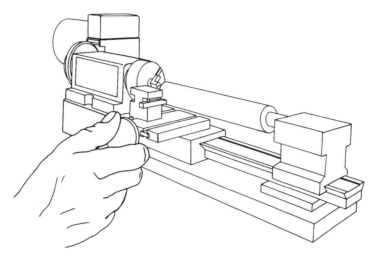

Measurements

THE PROCESS OF MAKING ACCURATE MEASURE-ments is probably the most important operation the new machinist must learn. Many of the parts you produce in your shop will not require critical measurement. When making small parts, however, you must be able to make accurate measurements to determine how much material must be removed to make the parts fit together properly. Rules, squares, calipers, and micrometers are tools that will be found in almost any machine shop. Your skill as a machinist will depend on your ability to use them.

STEEL RULES

The steel rule is the simplest of all measurement devices. It is used primarily to make rough measurements. Steel rules are available in a number of a lengths—the most common are 6, 9, and 12 inches long. They are also available with several different styles of gradations. Some rules are graduated in fractional increments, some in decimal increments, and some have both.

Because most machine work is dimensioned in decimals, most machinists prefer scales with decimal graduations, but decimal rules may not always be the most accurate to use. The smallest division found on a fractional rule is 1/64 of an inch. The smallest division on a decimal rule is usually 0.020 inch. Because the decimal equivalent of 1/64 is 0.015, the fractional rule has the finest divisions and is therefore more accurate. The final choice of rule should really be dependent on the dimension you are trying to read. If the dimensions you are trying to read is the decimal equivalent of a fraction, the fractional rule will be the most accurate one to use. If the dimension is a decimal that is not a fractional equivalent, the decimal rule will probably be most accurate. In any case, the rule cannot be used to make accurate measurements smaller than its finest divisions. To make more accurate measurements, you need calipers and micrometers.

SQUARES

Several different types of squares can generally be found in a machine shop. They are used for checking surfaces for right angle (90°) squareness, layout work, and making rough linear measurements. The most common square found in shops is the *combination square* (Figs. 2-1 and 2-2), which consists of a 12-inch steel rule and a movable head. The head has two ground surfaces, one at 90° to the rule and one at 45° to the rule. The head is movable and can be positioned and clamped in place at any point along the rule. These rules are normally graduated in fractions of an inch with the smallest division being 1/64 inch. They usually have a bubble level on the head and a small scriber stored in the head.

In addition to the head just described, two other types of heads are available on combination squares. One, a *protractor head*, has a ground surface that can be rotated to various positions to measure any angle from 0° to 180°, usually with an accuracy of 1°. The other head is a *centering head*. It has two ground surfaces at 90° to each other, forming a V with the apex of the V directly over one edge of the rule. This head is particularly useful in locating center on round stock.

Precision squares (Fig. 2-3) are hardened steel squares with fixed heads. As with most tools, they come in a variety of sizes with blade lengths ranging from 1½ inches to 36 inches. Normally the blades are not graduated. Because these squares have a fixed head and the working surfaces are hardened and ground, they are extremely accurate. They are used primarily in making precision setups and as an aid in adjusting and calibrating machines and shop equipment.

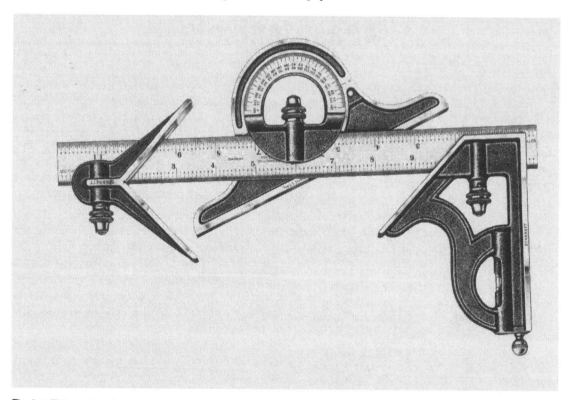

Fig. 2-1. This combination square consists of a 12-inch steel scale and three removable heads. The head at the right end of the square is used for establishing 45° and 90° angles. The center head is a protractor head and may be set to any angle from 0° to 180°. The remaining head is used for finding center on bar stock (courtesy of L. S. Starrett Co., Athol, MA).

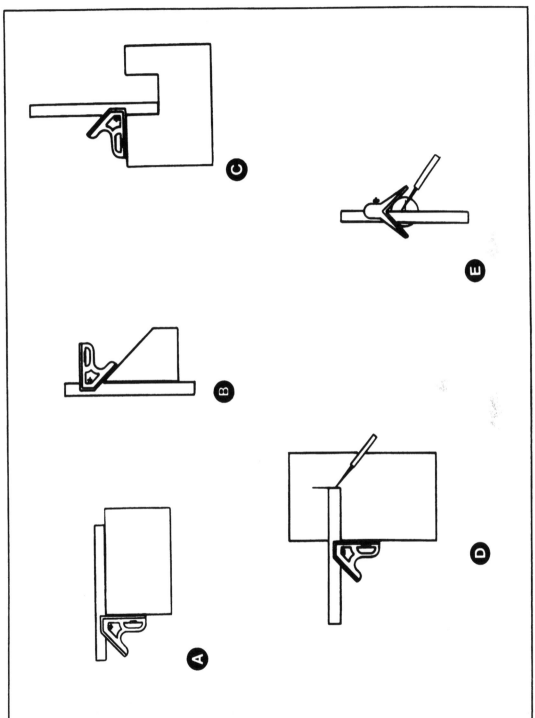

Fig. 2-2. A combination square can be used in many ways: (A) to check a 90° corner, (B) to check a 45° angle, (C) as a depth gauge, (D) to make a layout, (E) to find center on a piece of bar stock.

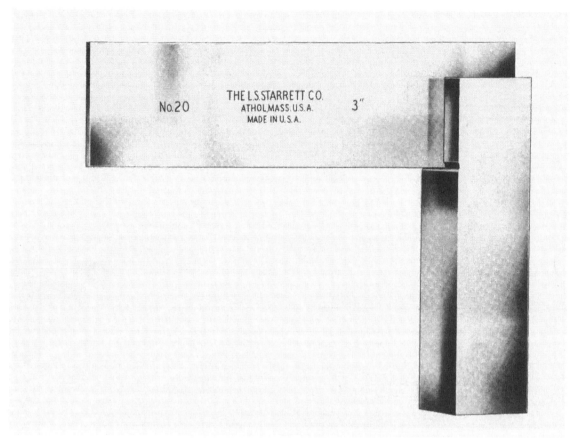

Fig. 2-3. Precision squares are used when extreme accuracy is required. They are made from hardened steel and are precision ground to ensure the highest degree of accuracy (courtesy of L. S. Starrett Co., Athol, MA).

SIMPLE CALIPERS

Simple calipers are not graduated and are only used to transfer measurements from some standard, such as a scale or an existing workpiece, to the part being made. Three common types are the inside caliper, the outside caliper, and the hermaphrodite caliper. All three are available in a number of sizes.

Inside and Outside Calipers

As their names imply *inside calipers* (Fig. 2-4) are used to measure inside dimensions such as the inside diameter of a tube, and *outside calipers* (Fig. 2-5) are used for measuring outside dimensions

such as the outside diameter of a tube or a piece of bar stock. Because the arms on calipers are made of fairly light metal, they may be sprung out of shape quite easily. They require a light touch. The accuracy achieved in making measurements with simple calipers is very much dependent on the machinist.

When taking measurements with simple calipers, adjust the jaws of the caliper to just slip over, or between, the surfaces being measured. You should feel just a slight amount of drag. Always recheck your measurement at least once to make sure the caliper setting hasn't slipped during the measurement process. Continued practice is required to develop the proper touch and get consistent measurements.

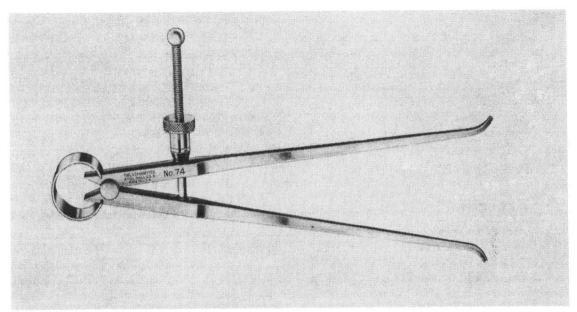

Fig. 2-4. The inside caliper is designed for transferring or comparing inside dimensions, such as the inside diameter of a piece of tubing (courtesy of L. S. Starrett Co., Athol, MA).

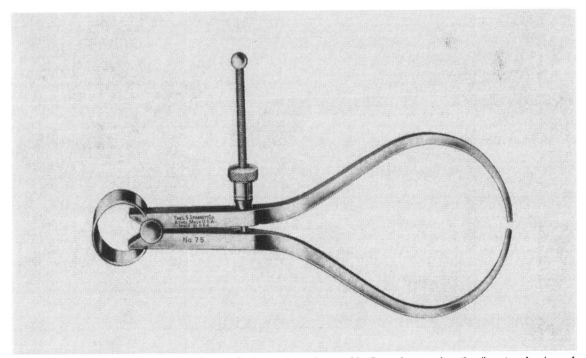

Fig. 2-5. The outside caliper is designed for transferring or comparing outside dimensions, such as the diameter of a piece of bar stock (courtesy of L. S. Starrett Co., Athol, MA).

Hermaphrodite Calipers

The *hermaphrodite caliper* (Fig. 2-6) is used primarily for layout work and for finding center on bar stock. When using the hermaphrodite caliper, paint the workpiece with layout ink. Then, use just enough pressure to scratch through the ink when scribing layout lines. This will reduce the wear on the caliper, and there will be less chance of the caliper setting slipping. Another benefit is that there won't be unsightly scribe lines left on your part when it is finished.

VERNIER CALIPERS

Unlike simple calipers, the *vernier caliper* is a true measuring instrument. It can be used for measuring inside dimensions, outside dimensions, and depths, all with an accuracy of .001 inch over the full range of the caliper (Fig. 2-7).

Most vernier calipers are made with separate inside measuring jaws, outside measuring jaws, and probe for making depth measurements. To use the caliper, simply select the proper set of jaws and slide the movable jaw until both jaws are just snugged up to the surfaces to be measured. Never use excessive force. The dimension is read from both the main scale and the vernier scale (Fig. 2-8).

The main scale is graduated in 1-inch divisions. 1/10-inch divisions, and 0.050-inch divisions. The vernier scale is divided into 50 divisions, each representing 0.001 inch. To read the caliper, first read from the main scale. To that reading, add the reading from the vernier scale. The reading on the main scale is the largest dimension readable between the zero on the main scale and the zero on the

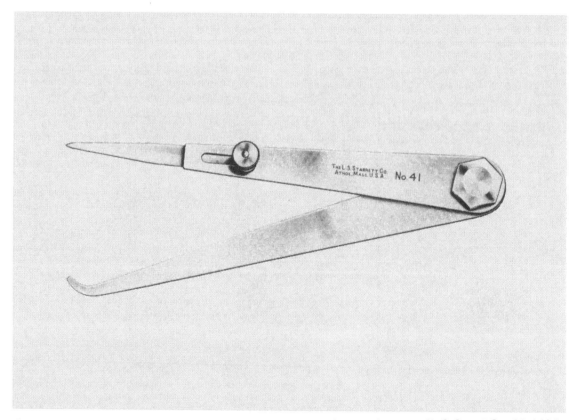

Fig. 2-6. The hermaphrodite caliper is used to establish edge to center distances (courtesy of L. S. Starrett Co., Athol, Ma).

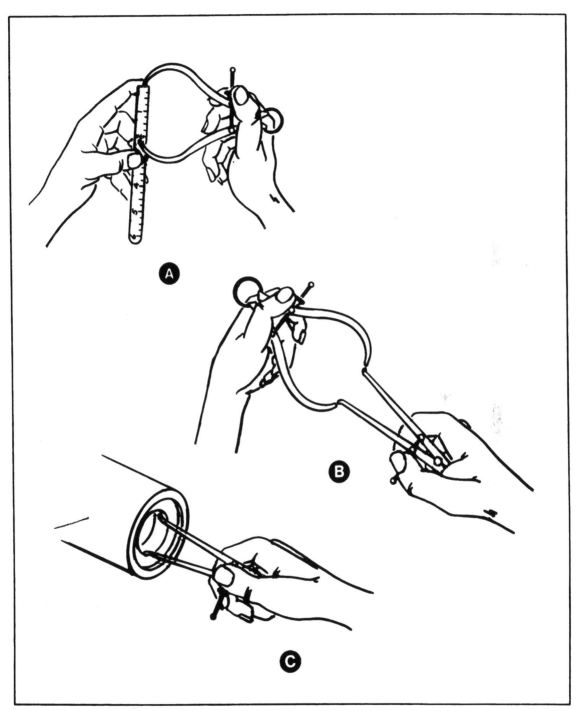

Fig. 2-7. A few of the many uses for calipers: (A) setting an outside caliper with a 6-inch scale, (B) transferring a measurement from an outside caliper to an inside caliper, (C) checking the inside diameter of a part.

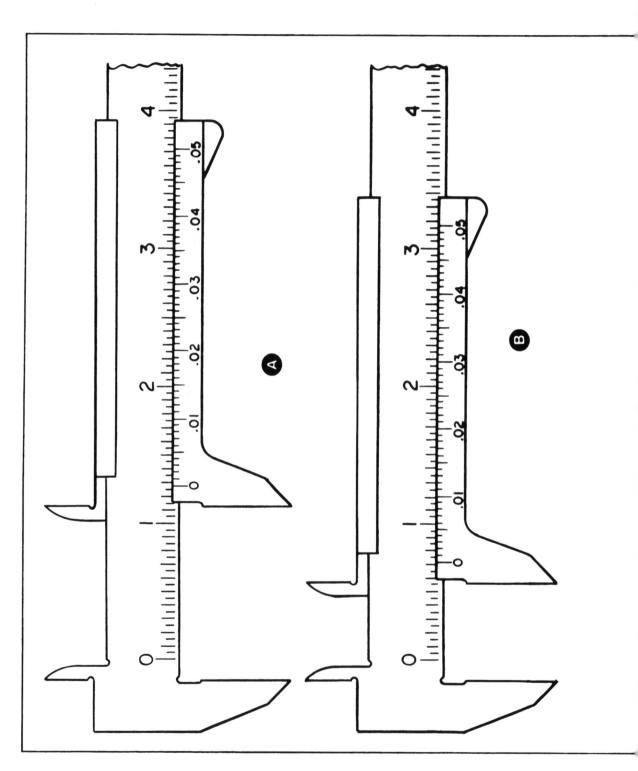

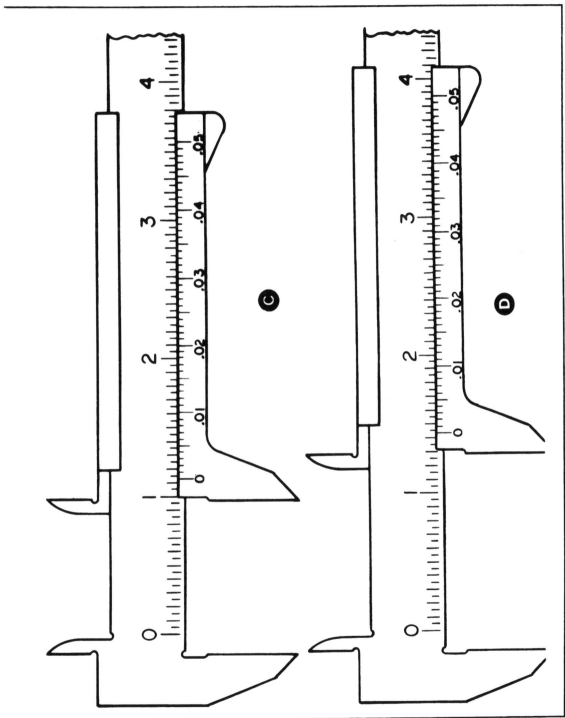

Fig. 2-8. Reading a vernier caliper is fairly simple once you master the basic principles. The readings on the four calipers shown above are: (A) 1.275 inches, (B) 0.712 inches, (C) 1.119 inches, and (D) 1.427 inches.

vernier scale. This will give you the dimension to the next smallest 0.050 division. The reading on the vernier scale is made by counting the number of divisions between the zero and the first line on the vernier scale that lines up with a line on the main scale. This will be the number of thousandths of an inch that you will add to the first reading to get the total reading.

As an example, note that in Fig. 2-8A the zero line on the vernier scale has passed the 1-inch mark, two 1/10-inch marks and a 0.050 mark. This means that the main scale is reading 1.250 inches. Next, reading on the vernier scale, note that the twenty-

fifth line on the vernier scale lines up exactly with a line on the main scale. By adding 0.025 to 1.250, you get a total reading of 1.275 inches. Check the readings on the calipers illustrated in Fig. 2-8 to see if you can read the correct dimensions.

DIAL CALIPERS

Dial calipers (Fig. 2-9) are similar to vernier calipers and can be used to make measurements to 0.001 inch also. The main advantage of the dial caliper is that it is easier to read than the vernier caliper. Inches and tenths of an inch are indicated on the main scale. Thousandths of an inch are indicated

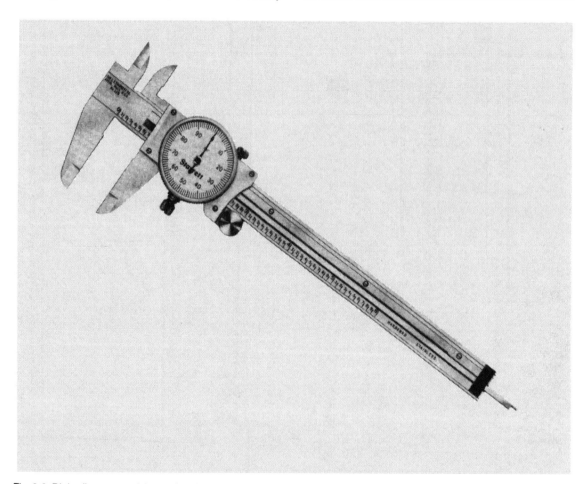

Fig. 2-9. Dial calipers are quicker and easier to use than a vernier caliper and can be read with greater accuracy (courtesy of L. S. Starrett Co., Athol, MA).

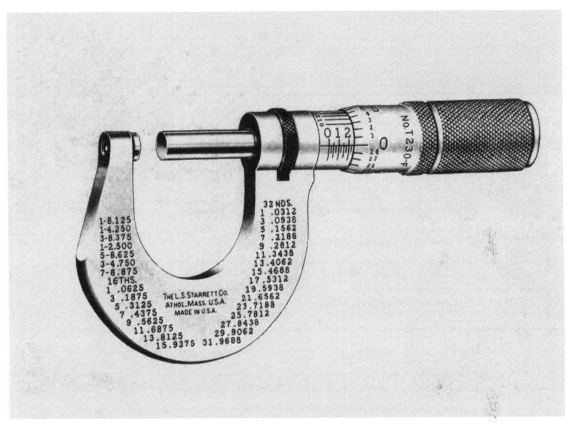

Fig. 2-10. The micrometer provides the most accurate method of making measurements. The 1-inch micrometer shown can be read to within .0001 inches (courtesy of L. S. Starrett Co., Athol, MA).

on the dial. To get a full reading, simply add the two together.

Most dial calipers are made so that the dial can be rotated for calibration. When using a dial caliper, always check to make sure the dial reads zero when the caliper jaws are closed.

MICROMETERS

The *micrometer* (Fig. 2-10) is a very accurate measuring device. Some micrometers are capable of making measurements to within 0.0001 (one ten thousandth) of an inch. Because of this accuracy, the measurement range is generally restricted to one inch. Therefore, micrometers are made in a number of sizes. The most commonly seen size has a range of 0 to 1.000 inch.

The accuracy of a micrometer is dependent on a lead screw. Better micrometers all have ground and polished lead screws rather than rolled lead screws (Fig. 2-11). The lead screw has 40 threads per inch so that every revolution of the screw advances the spindle by twenty-five thousandths.

To use the micrometer, hold the frame in the left hand. Revolve the thimble until the part to be measured is just snugged up between the anvil and the spindle. Make sure the micrometer is perpendicular to the surfaces being measured. If possible, it is best to read the micrometer before removing it from the workpiece. Do not clamp the micrometer tightly onto the workpiece. Excessive force can damage the lead screw.

How to read the micrometer is illustrated in

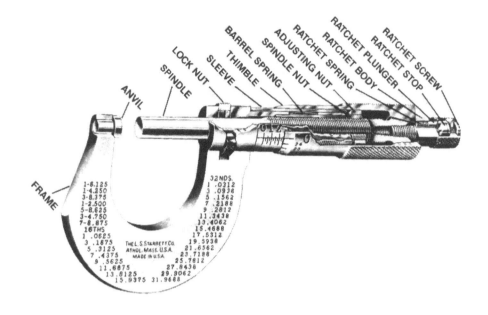

Fig. 2-11. This cutaway illustration shows the internal details of a micrometer and identifies its various parts. The best micrometers have precision ground screw threads for the highest degree of accuracy (courtesy of L. S. Starrett Co., Athol, MA).

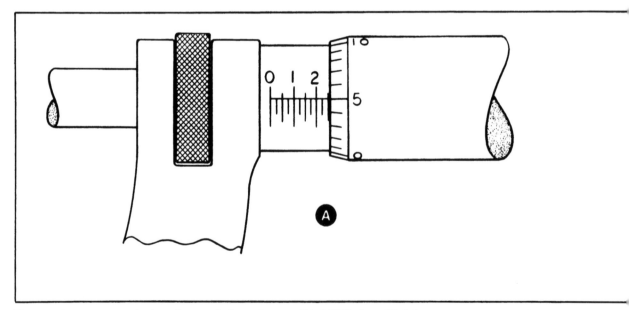

Fig. 2-12. The readings on the three illustrated micrometers are: (A) 0.255 inches, (B) 0.775 inches, and (C) 0.432 inches.

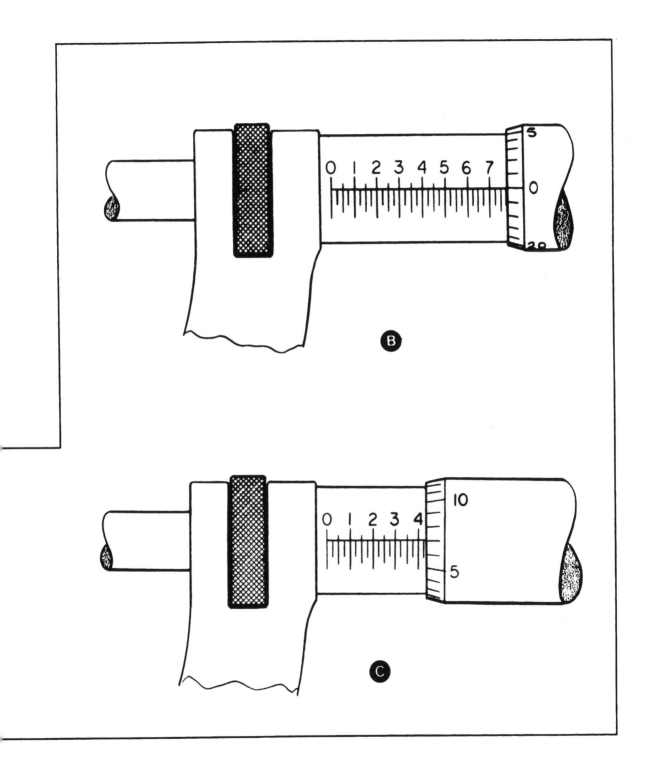

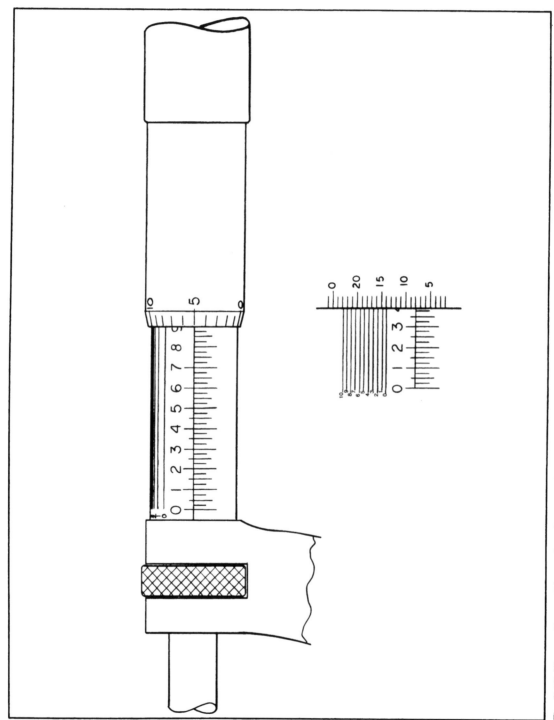

Fig. 2-13. The reading on this vernier micrometer is 0.905 inches. The reading on the vernier scale inset is 0.3829.

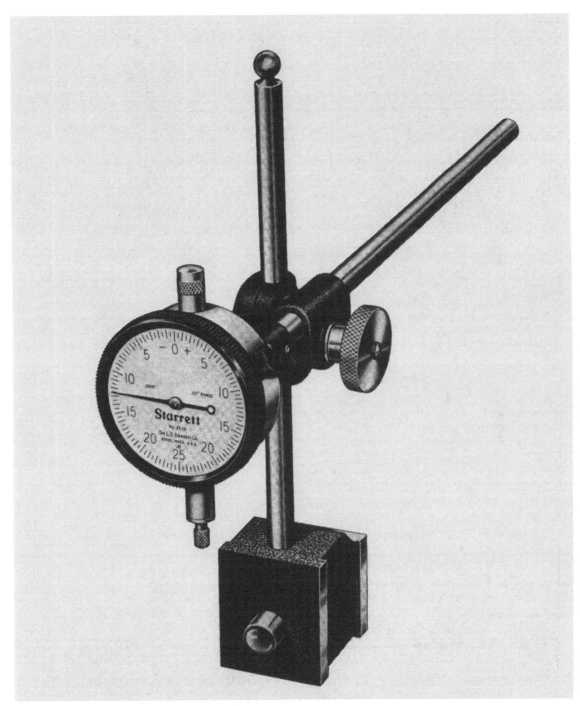

Fig. 2-14. Dial indicators are used for setting up lathes, milling machines, and other machine tools. This indicator is mounted on a magnetic base (courtesy of L. S. Starrett Co., Athol, MA).

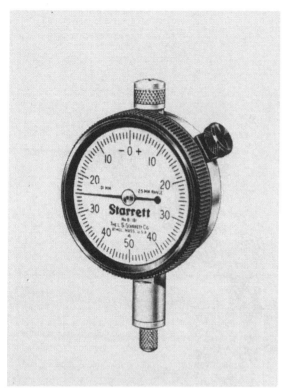

Fig. 2-15. A balanced dial indicator reads in both directions. This one has metric calibration and is graduated in 0.01-mm increments (courtesy of L. S. Starrett Co., Athol, MA).

Fig. 2-12. The sleeve of the micrometer is graduated in tenths of an inch (0.1), which are then divided into four parts or 0.025-inch increments. the 0.1-inch divisions are numbered consecutively up to 10. The beveled edge of the thimble is divided into 25 divisions, each representing 0.001-inch. Every fifth division line is longer and numbered 0, 5, 10, 15, and 20. To read the micrometer, first take the largest reading visible on the sleeve and then add to it the number of thousands (0.001 inch) indicated on the thimble.

The micrometer illustrated in part A of Fig. 2-12 is set to read 0.255 inches. The largest number that can be read on the sleeve is 0.250. To this, you add the number that can be read on the sleeve is 0.250. To this, you add the number of thousandths indicated on the thimble or 0.005. The combined reading is then 0.250 + 0.005 or 0.255 inches.

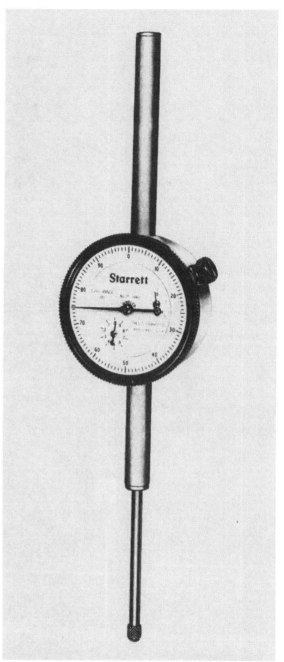

Fig. 2-16. A continuous dial indicator reads in one direction only. This one has a range of 2.000 inches and is graduated in 0.001-inch increments. The small left-hand dial is graduated in 0.01-inch increments, and the small right-hand dial indicates inches (courtesy of L. S. Starrett Co., Athol, MA).

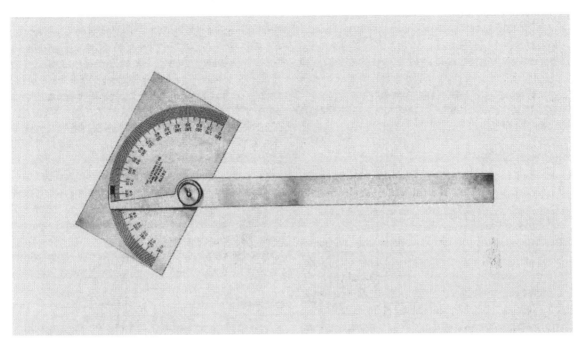

Fig. 2-17. Protractors are used to measure angles. This one reads from 0° to 180° and is graduated in 1° increments (courtesy of L. S. Starrett Co., Athol, MA.).

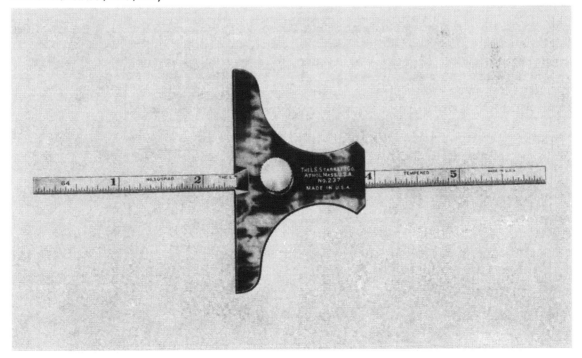

Fig. 2-18. Depth gauge (courtesy of L. S. Starrett Co., Athol, MA).

For micrometers capable of reading tenths of thousandths (0.0001 inch), a vernier scale is added. This type of micrometer is illustrated in Fig. 2-13. The micrometer is read in the same way as a standard micrometer, then the number of tenths of thousandths indicated on the vernier scale is added to the reading.

DIAL INDICATORS

Dial indicators (Fig. 2-14) are sensitive devices that can be used in many ways. Their greatest value will be in making precision setups on lathes and milling machines. They will be used to check the alignment of the work piece to the machine.

Dial indicators are made with a variety of different dial styles. The two most common are the balanced dial and the continuous dial. The *balanced dial* (Fig. 2-15) reads in two directions and gives plus or minus measurements. The *continuous dial*, (Fig. 2-16) reads in one direction only, usually clockwise starting from zero.

The dials are usually graduated in thousandths of an inch (0.001 inch), but some of the more accurate ones may be graduated in tenths of thousandths of an inch (0.0001 inch). The range of an indicator may be as little as 0.015 inch or as much as 1 inch depending on the type of use for which it was intended. Indicators that have a large range usually have two dials, a small dial indicating tenths of an inch (0.1 inch) and the larger one indicating thousandths (0.001 inch) or tenths of thousandths of an inch (0.0001 inch).

Some dial indicators are equipped with interchangeable contact points and may have three or four different types of contact points. The type of point used depends on the application and the preference of the machinist. There are also a variety of different styles of indicator holders or supports available. Some have magnetic bases while others use a mechanical clamp of some sort. All are equipped with a variety of support rods and universal clamps to make them as versatile as possible.

CAUTION

Never take measurements on any machine while it is running. Always shut the power off and wait until the machine comes to a complete stop.

All measurement devices should be treated as precision instruments. They should be kept clean and lightly oiled. Whenever possible, they should be stored in dust-proof boxes or in a clean drawer.

Chapter 3

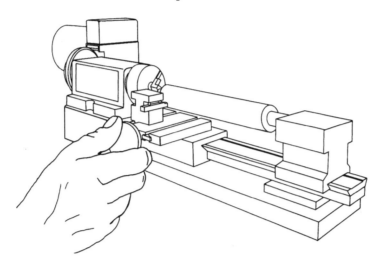

Hand Tools

SCRIBERS, PUNCHES, HAMMERS, MALLETS, FILES, and saws are all small tools essential to the operation of a small machine shop. It might seem that their use is so simple that there wouldn't be much to say about them. Actually, complete books have been written about these mainstays of the machinist's craft. Each of these tools has evolved over many years of use, and many variations have been developed to fill specific needs. In this chapter I will discuss some of the more common variations of these tools and their intended use.

SCRIBERS

Scribers (Figs. 3-1 and 3-2) are used to mark materials and make layouts. Unlike pencils, crayons, and pens, they mark by scratching or engraving a mark on the material. Because of this, they have to be made of something much harder than the material on which they will mark. Most scribers are made of high-carbon tool steel or tungsten carbide. They are also tempered for

maximum hardness. A good scriber can be used to mark on most metals, even unhardened tool steel, but rather than use a scriber to scratch lines into hard materials, most machinists prefer to coat the part with layout ink. The scribe is then used to scratch lines in the ink.

Layout ink is a fast drying ink, usually dark red or purple, that is painted directly onto the part being made (Fig. 3-3). When it dries, it leaves a thin coating on the part that is easily cut through by the scribe. The use of layout ink has several advantages. First, the resulting scribe lines are easier to see than lines scratched into the metal (Fig. 3-4). Second, there is less wear on the point of the scriber so the scriber will last longer. Third, when you finish the part, it is a lot easier to remove the layout ink by washing the part in a solvent than it is to polish out lines scribed into the part (Fig. 3-5).

There are several safety rules to be remembered when using a scriber. First, never use the scriber as a punch or as a pry bar. The point has been heat-treated for maximum hardness and is

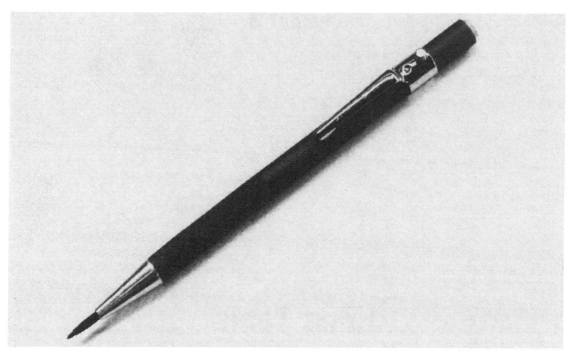
Fig. 3-1. Scribers are used to scribe layout lines on a part.

Fig. 3-2. Scribers like this one have a retractable point so it is safe to carry in your pocket.

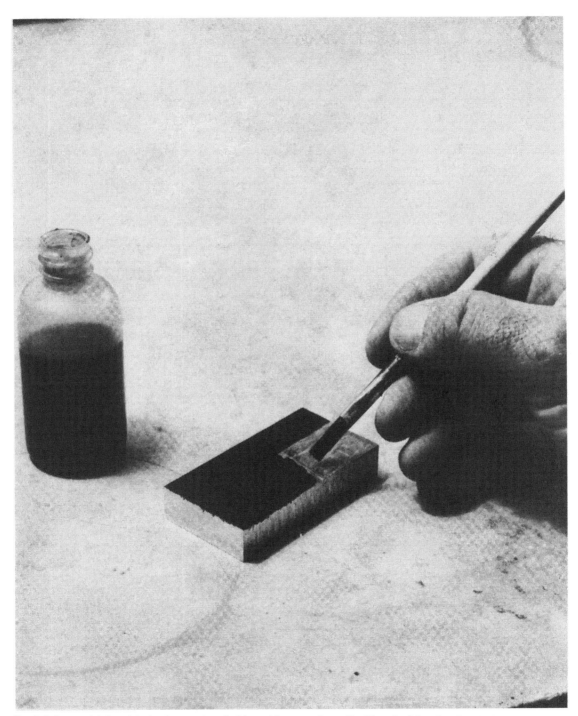

Fig. 3-3. Layout ink is painted onto a part so that layout lines can be scribed into the ink rather than into the part.

Fig. 3-4. The layout is easily scratched into the ink with a scriber. The resulting lines are not only easier to see, but they are also easier to remove when the part is finished.

Fig. 3-5. With the center lines clearly scribed onto the part, it is an easy task to center punch hole locations prior to drilling.

quite brittle. Should the point break off, it could be a very dangerous projectile. Second, never carry a scriber in your pocket unless it has a good protective cover over the point or points. (Some scribers have retractable points. If you want to carry a scriber in your pocket, these are probably the safest kind to carry.)

PUNCHES

There are four basic types of punches used in the machine shop: the prick punch, the center punch, the pin punch, and the drift punch.

Prick Punches and Center Punches. Prick punches and center punches (Fig. 3-6) are quite similar. Both are used to locate the center before drilling a hole. The center punch has its point ground to approximately 90°. The prick punch is ground with a sharper point, usually about 60°. Prick punches are used when doing extra fine layout work. The sharper point makes it easier to position accurately. Because the depression left by a center punch is better suited for starting a drill, the prick punch is usually used first, then a center punch is used to shape the depression. The use of a prick punch is not necessary when the location of a hole is not critical.

Sometimes the mark left by a center punch may not be located exactly where you want it. If this happens, it is usually quite easy to move the mark. Using a prick punch, mark the side of the center punch mark in the direction you want to move it. Then, carefully center punch it again (Fig. 3-8). The center punch will move in the direction of least resistance toward the prick punch marks. To get the mark exactly where you want it, you may have to repeat this procedure two or three times.

Spring-loaded center punches are real time savers. You simply locate the punch on the workpiece and push down. A spring release mechanism automatically trips inside the punch with sufficient force to mark the workpiece. This is accomplished without the use of a hammer. In the long run, the time saved by an automatic center punch makes it worth the investment.

Pin Punches and Drift Punches. The only difference between a pin punch and a drift punch is that pin punches are made straight, while drift punches are tapered. Pin punches are used in assembly and disassembly of parts held together with dowel pins. Drift punches are primarily intended to be used to align bolt holes when assembling parts. Both types of punches are made from high-grade alloy steel, heat-treated for maximum toughness. They come in a variety of sizes, ranging from about 1/16-inch in diameter and up (Fig. 3-9).

After repeated usage, the head of any tool that is hammered on may mushroom over (Fig. 3-10). When this happens, you should always redress the

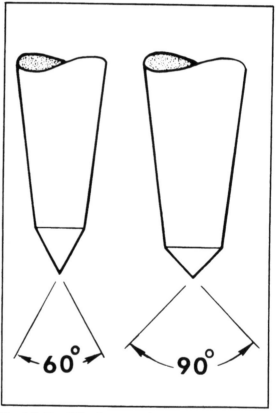

Fig. 3-6. The 60° point on a prick punch permits greater accuracy in locating a punch mark. The 90° point on a center punch does a better job of centering a drill bit. For best results, use a prick punch to locate the punch mark and then use a center punch to reshape the mark for drilling.

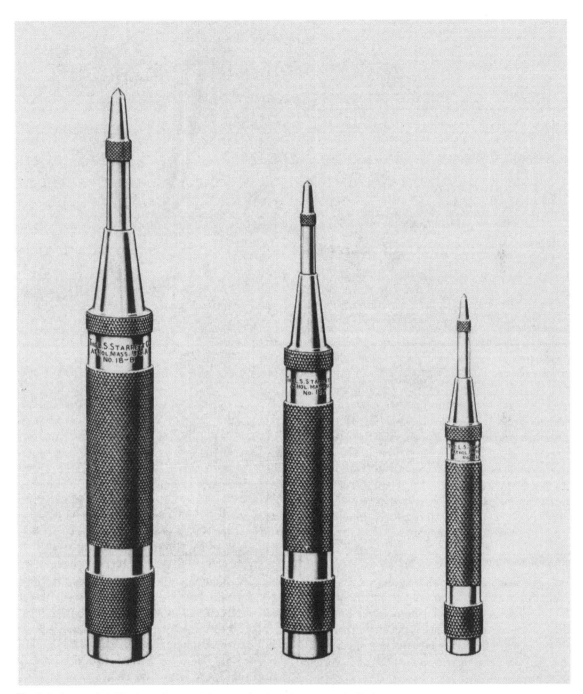

Fig. 3-7. Automatic center punches contain a mechanism that automatically impacts the punch when downward force is applied. Since it is not necessary to strike the punch with a hammer, both hands can be used to locate the punch and steady it (courtesy of L. S. Starrett Co., Athol, MA).

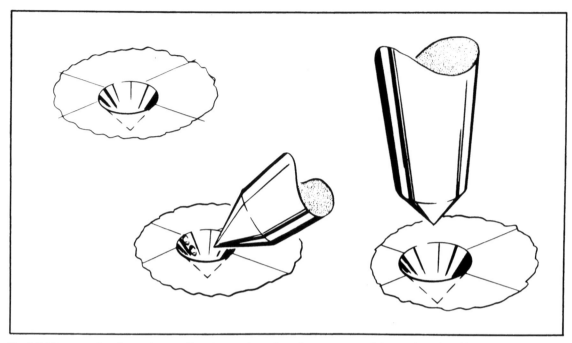

Fig. 3-8. The procedure for moving an off-center punch mark: make two or more light punch marks with a prick punch in the direction you want to move. Repunch the mark with a center punch. It may be necessary to repeat this process two or three times to place the mark exactly where you want it.

tool by grinding the mushroomed material away. After grinding, carefully inspect the tool for cracks. If you find any, the head of the tool should be ground down until the crack is completely removed to minimize the danger of small chips breaking off when the tool is struck. A good safety rule to observe is to always wear safety glasses when using a hammer to strike another tool.

HAMMERS AND MALLETS

The hammer is one of the oldest and simplest tools that we have. It probably comes in more different sizes and shapes than any other tool (Fig. 3-11). Over the years, man has developed hammers to fill many different needs. In this book I will discuss only those most commonly used in working with metals.

The method of holding and striking with a hammer is basically the same for all hammers. The only difference is that small, light hammers are usually swung using a wrist motion. Larger hammers are usually swung from the elbow. The hammer should always be gripped near the base of the handle because this allows the best control (Fig. 3-12). Hammers should never be struck on their side or on an edge. The face and the peen are the only areas of the hammer head that are hardened to withstand the shock of impact. Improper striking could result in chipping or breaking the tool.

Even with proper use, there is always the danger of chipping either the hammer or the workpiece. For this reason, it is always a good idea to wear safety glasses when using a hammer.

Ball Peen Hammers. The ball peen is probably the most common type of hammer found in a machine shop. They are available in a wide range of sizes, from those with 4-ounce heads and 10-inch handles, up to those with 32-ounce heads and 16-inch handles. They are generally forged from high-carbon steel and tempered for hardness and tough-

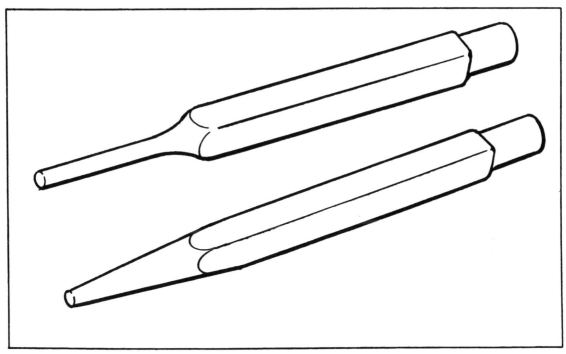

Fig. 3-9. This drawing illustrates the difference between a pin punch and a drift punch.

ness. The face is slightly crowned and usually has a generous bevel around it to help prevent chipping. The *peen* on a ball peen hammer is spherical in shape. It is used extensively in working metals.

Old world craftsmen used ball peen hammers to work harden metals such as the brass plates used in clocks. Brass and many other copper alloys are not hardened by heat treating; they are hardened by working with a hammer or by rolling. In the old days a clockmaker would lightly peen both sides of a piece of brass, and the resultant work-hardening made the material more suitable as a bearing material.

Ball peens are also used to form sheet metal. Repeated peening on one surface of a sheet of metal causes that surface to expand, and the expansion causes the sheet to become dish-shaped. This method of metal working was used extensively in hand forming copper utensils and jewelry.

Cross Peen Hammers. The cross peen is also a popular metal working hammer. It was used extensively by blacksmiths as a forging hammer. The peen is wedge shaped with the wedge running crossways to the axis of the handle. The point of the wedge is usually rounded with a generous radius. Like the ball peen, cross peen hammers are available in many sizes. Starting with 4-ounce heads, they range up to heavy-duty hammers with heads weighing 2 pounds or more. Handle lengths vary from about 10 inches on the smaller ones up to 16 inches on the heavier ones.

Riveting Hammers. The riveting hammer is quite similar to the cross peen hammer. Where the cross peen hammer has a rounded surface on the peen, the riveting hammer has a sharp chisel-like face. This chisel face is used in spreading rivets or in staking. As a rule, these hammers are quite light. The head weight ranges from 2 to 6 ounces. Handle length is usually about 10 inches.

Chasing Hammers. The chasing hammer or *silversmith's hammer*, as it is sometimes called, is a lightweight ball peen hammer with a special handle.

43

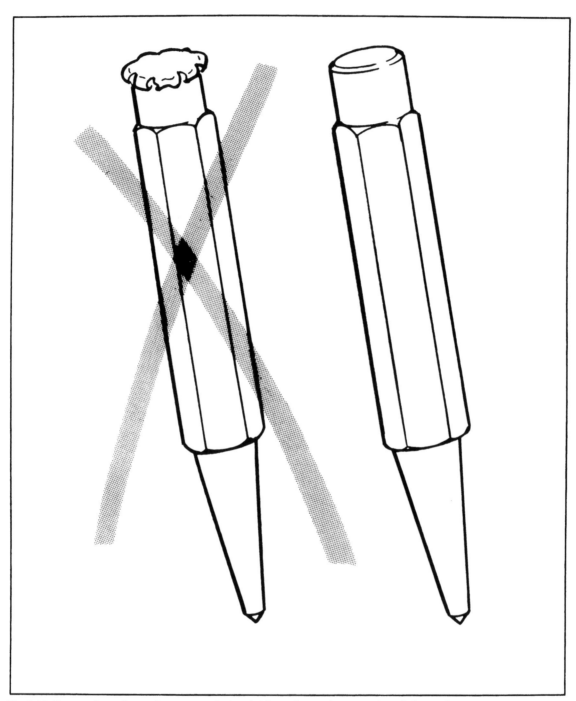

Fig. 3-10. Repeated use of a punch may cause the head of the tool to mushroom, and small pieces of metal may fracture off and cause an injury. Mushroomed tools should be carefully dressed on a grinder to eliminate the deformed material.

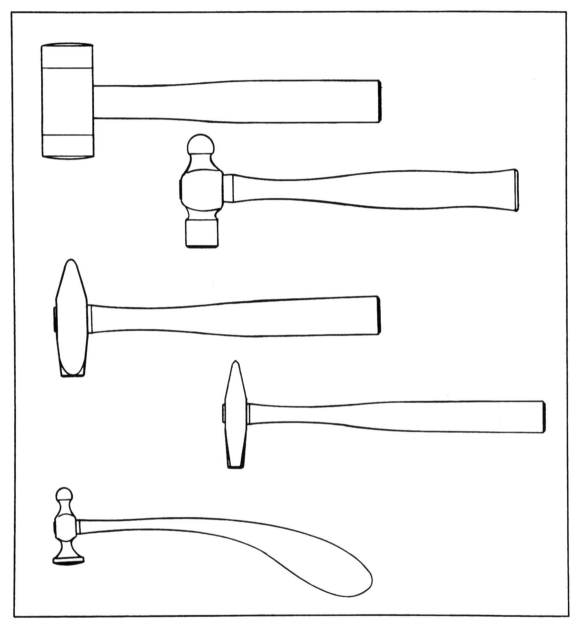

Fig. 3-11. There are many different kinds of hammers. From top to bottom; mallet, ball peen, cross peen, riveting, and silversmith's chasing hammer.

The head weight is usually 2 to 3 ounces. The face of the hammer is unusually large for a hammer this light. The handle is curved and is designed for working with the silversmith's forearm lying on the bench. The hammering motion comes from the wrist alone. Because the silversmith's work is so fine and delicate, this method of working gives him the required control.

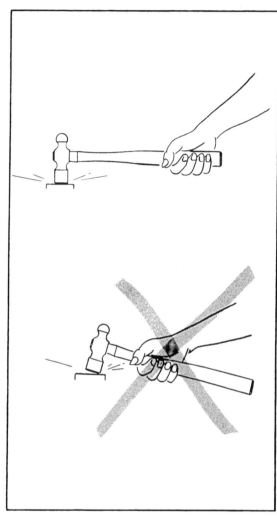

Fig. 3-12. For proper control, a hammer should always be gripped near the base of the handle. With all hammers except very light or very heavy ones, the swinging motion should come from the elbow. Never strike with the side or edge of the hammer.

Mallets. Mallets or *soft face hammers* are used when it is desirable to strike an object without marking it. As an example, when setting up a workpiece on a milling table, the machinist will lightly clamp the part to the table and then use a mallet to nudge it into the correct alignment. With the face of the mallet being much softer than the workpiece, there is little chance of damaging the workpiece.

Mallets are made from a variety of materials. Hardwood, rawhide, hard rubber, lead, brass, and plastic are some of the more common ones. Some mallets, especially those with plastic faces, are made so that the faces can be removed and replaced. The type of mallet used in a shop depends a great deal on personal preference.

FILES

Files rank with hammers as being one of the oldest tools known to man. The first files were probably rough rocks that were used as abrasives to

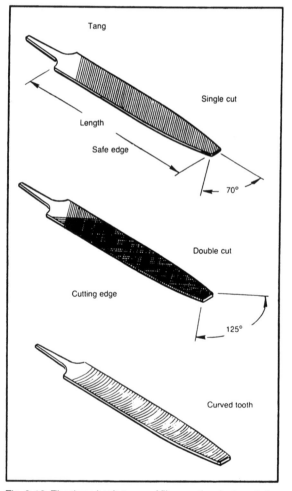

Fig. 3-13. The three basic types of files are the single-cut, the double-cut, and the rasp.

46

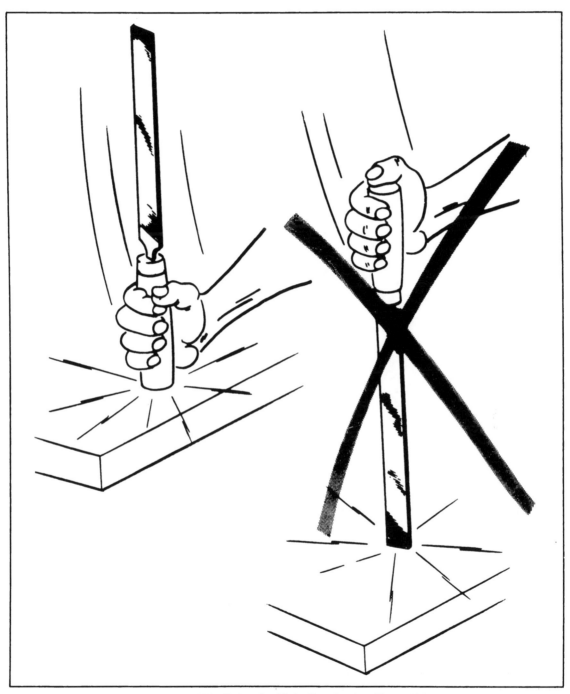

Fig. 3-14. To seat a file handle properly, hold the file handle with the file pointed up. Then, with a quick downward motion, strike the handle on your workbench. The momentum of the file will seat it into the handle. Never attempt to strike the file to force it into the handle.

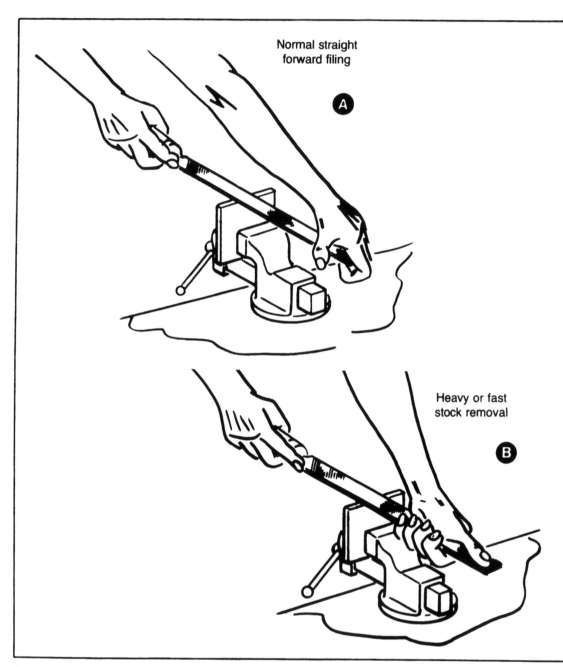

Fig. 3-15. The four basic methods of filing are shown here. (A) For normal filing the file should be gripped with both hands. A downward force is applied on the cutting stroke, and the file is lifted clear of the work on the return stroke. (B) For heavy stock removal with a coarse file, the file is gripped with heavier downward pressure and worked in a straight forward motion. (C) For fine finish filing less downward force is required, and the file is held to distribute force more evenly over the file. (D) Draw filing is used for finish work when a nice flat finish is required. The file is held with both hands and drawn straight toward the body with a light downward pressure.

48

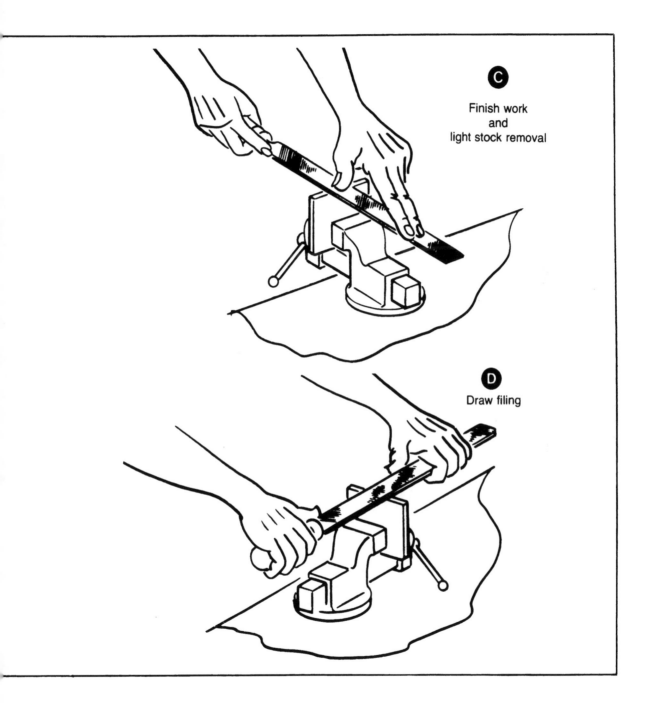

C

Finish work
and
light stock removal

D

Draw filing

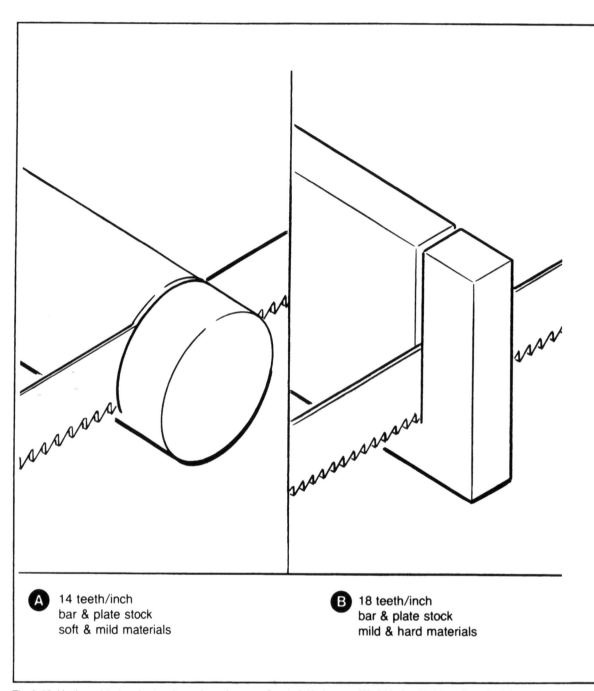

A 14 teeth/inch
bar & plate stock
soft & mild materials

B 18 teeth/inch
bar & plate stock
mild & hard materials

Fig. 3-16. Hacksaw blade selection depends on the type of material being cut. (A) A blade with 14 teeth per inch is used to cut soft bar and plate stock. (B) A blade with 18 teeth per inch is used for mild steel plate, bar, or heavy structural shapes. (C) A blade with 24 teeth per inch is used for hard bar or plate stock, aluminum extrusions, and light structural shapes. (D) A blade with 32 or more teeth per inch is used for cutting small extrusions and thin wall tubing.

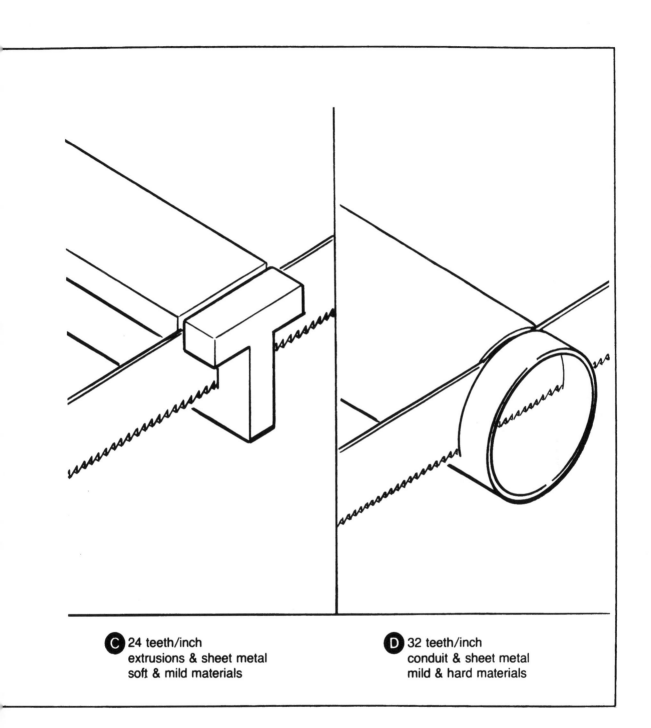

C 24 teeth/inch
extrusions & sheet metal
soft & mild materials

D 32 teeth/inch
conduit & sheet metal
mild & hard materials

shape other rocks. Centuries later, at the start of the iron age, files were handcrafted by cutting a series of teeth into an iron bar with a hammer and chisel. Each row of teeth was cut individually. (The first job given to an apprentice metal worker or clock maker usually was to handcraft his own set of tools, which always included a number of files.)

Files as we know them today became possible around the middle of the eighteenth century, when the first successful file cutting machine was invented. Today, there are many different styles of files available to the craftsman. They come in a variety of sizes, shapes, and tooth styles (Fig. 3-13).

Single-cut files have parallel rows of teeth cut at approximately 70° to the center line of the file. *Double-cut files* have rows of teeth cut crisscrossed. One set of parallel cuts at approximately 70° to the center line of the file, and the other set at approximately 125° to the first set.

Files also come in varying degrees of coarseness, starting with smooth, then second cut, bastard cut, and finally coarse cut. The degree of coarseness is comparable only when files of the same size and shape are compared. A 6-inch smooth file has more teeth per inch than a 12-inch smooth file. The length of a file is usually measured from the heel to the point, not including the tang. The exception to this rule is the Swiss file, which is measured length overall including the handle.

Filing Techniques

Good filing is an art. In the old days, craftsmen were frequently judged by their ability to file (Fig. 3-15). One story claims that a well-known prestigious car manufacturer would not hire a machinist until he had successfully demonstrated his filing skill. He was given an iron bar and an iron plate. When he was finished, the bar had to be filed square. The plate had to have a square hole filed through it. If the square bar did not fit perfectly into the square hole no matter how it was turned, the machinist did not get the job.

Work to be filed should be clamped firmly to the workbench or held in a vise. The work should be approximately waist high for most jobs. The area in front of the workbench should be kept clear and unobstructed. If some of the surfaces of the workpiece are already finished, they should be protected by placing a strip of some softer material between them and the vise. Copper, brass, aluminum, wood, and leather are materials frequently used to protect finished surfaces.

There are three basic techniques used in filing. The first is *straight forward filing*. Used for roughing and semifinishing operations, this technique is accomplished by pushing the file straight across the work. A slight downward pressure is maintained on each cutting stroke, and the file is lifted clear of the work on each return stroke.

The second method is *draw filing*. Draw filing is accomplished by holding the file crossways or perpendicular to the direction of the stroke. Even downward pressure is maintained on both the cutting stroke and the return stroke. This method of filing is used when a smooth, level finish is desired.

The last method is *lathe filing*. As the name implies, it is accomplished while the workpiece is in the lathe. This method is frequently used to correct slight tapers or bows that sometimes occur when

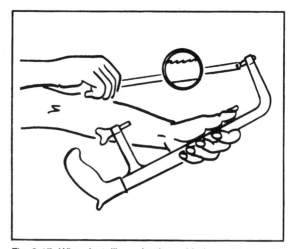

Fig. 3-17. When installing a hacksaw blade, make sure the teeth point away from the handle. The installed blade should be tight enough to hum when tapped sharply on the side.

the workpiece is long and slender and deflects under the pressure of the cutting tool. When lathe filing, the turning speed should be approximately 50 percent faster than the normal turning speed. Single-cut files are usually used for this operation. Hold the file at a slight angle and use long steady strokes, maintaining a uniform downward pressure. Check the file frequently to make sure it is not clogging or pinning (Fig. 3-15). *Caution:* never lathe file with a file that does not have a good handle.

File Care

A file should be cleaned whenever it starts to clog (or *pin*). This is accomplished by brushing the file with a *file card*, a special brush with short steel bristles. Stubborn particles may have to be removed with a pick. There is no good way to prevent a file from clogging, but one trick that sometimes works is to rub a small amount of chalk on the file.

Always keep your files clean. They should be cleaned after every use and whenever they become clogged. Files should also be kept away from water, which will cause them to rust, and oil, which lubricates them and causes them to slide over the work and not cut. When you store your files, wrap them in paper or cloth to prevent damage from other heavy tools stored with them. Files should always be fitted with handles. The handles will make the file easier to use and control and safer to use.

SAWS

Hacksaws are used in the shop to cut metals. They consist of a frame, a handle, and a blade. The frames are usually adjustable to take several different blade sizes. The pins that hold the blade can be rotated so that the blade can be set at several different angles to the frame.

Blades are usually 8 to 12 inches long, approximately ½ inch wide and 0.025 inch thick. Most of the blades are made of high-carbon steel, but some of the more expensive ones may be made of high speed tool steel. These blades last longer and are better suited for cutting tougher metals. Blades are classified by the number of teeth per inch. Coarser blades are used for soft bar stock and the finer blades are used for hard metals, sheet metals, and tubing. For a quick blade selection guide, refer to Fig. 3-16.

When installing a blade into the frame, the cutting edge of the teeth should point away from the handle (Fig. 3-17). The blade should be tightened in the frame until it hums when tapped sharply on the side. The material to be cut should be clamped tightly in a vise with the vise as close to the cut area as possible. Sheet metal can be clamped between two blocks of wood for support.

Use your left thumb to guide the blade when starting a cut (Fig. 3-18). After the cut is started, grip the saw with both hands, one hand on the handle and the other at the opposite end of the frame. Apply a light downward pressure on the cutting stroke and release the downward pressure on the return stroke. Your strokes should use the full length of the blade. Don't attempt to stroke too

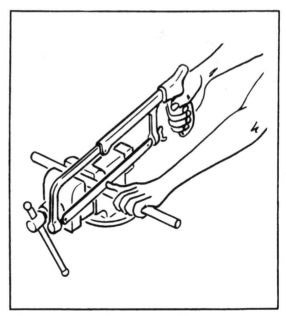

Fig. 3-18. Use your left thumb as a guide to start a cut. Once the cut is started, grip the saw with both hands. A light downward pressure is used on the cutting stroke. Too much pressure will cause the blade to turn.

fast or use too much downward pressure. This results in broken or bent blades. When the cut is almost completed, ease up on the downward pressure. The free end of the material can be supported by your left hand to prevent it from falling on the floor.

Chapter 4

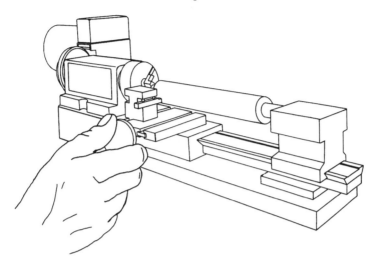

Bench Tools

V ISES, BENCH GRINDERS, DRILL MOTORS, and drill presses are the tools most commonly thought of as bench tools. In this chapter I will discuss each tool and its related equipment. I will outline a few of the safety rules that should be observed when using them.

VISES

Vises are used to hold parts while you are working on them. For filing, drilling, and other shop operations, the vise is practically indispensable. A good heavy vise with 3½- or 4-inch jaws, bolted securely to the workbench, is ideal for holding a part while filing or drilling it with a hand-held drill motor. Smaller vises with vacuum bases or clamp-on bases are ideal for holding small parts while doing finishing operations. Machine vises are used with drill presses and milling machines.

Bench Vises. Bench vises (Fig. 4-1) are usually made from cast iron. Most have removable serrated jaw plates that can be replaced when they become worn. Some have a swivel base to allow you

to position your work for the greatest advantage. As another feature, many of these vises are built to double as an anvil. Bench vises come in many sizes. I have found that one with 3½-inch-wide jaws will handle most of the projects I want to tackle.

When clamping work with finished surfaces, cover the jaws with a soft material to prevent damaging the surfaces. Jaw liners are easily fabricated from a piece of soft copper or brass sheet stock. The material should be approximately 0.020 inch thick. Simply cut out a shape as shown in Fig. 4-2 and bend it to fit over the jaws. If the job is properly done, the liners should lift off for storage when they are not required.

Portable Vises. Small portable vises are handy for many applications. These vises usually have jaws 1½ to 2 inches wide. Most of these vises are made to clamp onto a table top. Some are made with a vacuum base made from a large rubber suction pad that is actuated by a lever (Fig. 4-3). Just position it where you want it on the bench top and raise the lever to secure it in place. Portable vises

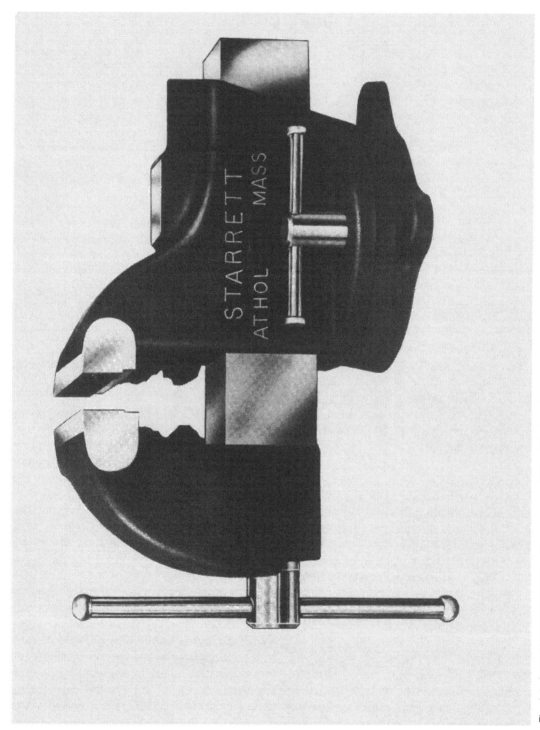

Fig. 4-1. A good bench vise is a must in every shop. It should be heavily constructed and mounted rigidly to your workbench. A vise with 3½- to 4-inch wide jaws is ideal for most work. (courtesy of L. S. Starrett Co., Athol, MA).

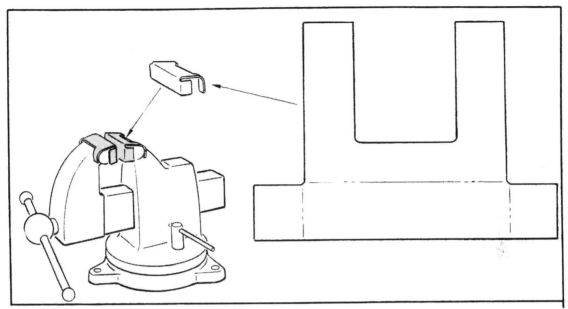

Fig. 4-2. Removable jaw liners are easily made from soft aluminum sheet stock, preferably about 0.02 inches thick. These liners will prevent marring the finish on nearly completed projects.

Fig. 4-3. A vacuum base vise can be used anywhere there is a clean, smooth surface. For those without a permanently setup shop, these vises are indispensable.

are especially handy in a small workshop where the vise might be in the way if mounted permanently.

Machine Vises. Drill vises and milling machine vises (Fig. 4-4) are special vises. The bases and sides are machined perpendicular to the jaw surfaces so that the workpiece will be held true to the axis of the drill or cutter. The jaws in these vises are usually smooth so they can be used safely on finished surfaces. Sometimes they have V grooves milled into one face for holding round bar stock. Most of these vises have provisions for clamping them onto the machine.

BENCH GRINDERS

Bench grinders (Fig. 4-5) are used primarily for sharpening tools. Most bench grinders have two wheels—one coarse and one fine. If you use carbide tools in your shop, you may prefer to have one wheel for sharpening tools made of tool steel and one wheel for sharpening carbide tools.

There should be a tool rest in front of each wheel. If the tool rest is adjustable, it should be moved in as close to the wheel as possible without actually touching it. This will prevent your workpiece from accidentally becoming wedged between the wheel and the tool rest. After adjusting the tool rest, always rotate the wheel through at least one revolution by hand to make sure it is not slightly out of round. If the wheel is out of round, it should be dressed.

Fig. 4-4. The machine vise is a portable vise for holding small parts during drilling, reaming, and light milling operations. To simplify setup procedures, both the sides and the base are carefully machined square to the jaws.

Fig. 4-5. Bench grinders are used for shaping and sharpening tools. A good grinder should have a speed of approximately 3400 rpm Work rests, safety shields, and covered wheels are all necessary for safety reasons.

Truing or *dressing* a wheel serves several purposes. It not only corrects the slightly out-of-round condition, but sharpens the wheel by removing metal particles that clog the wheel, and by fracturing away the dull, worn abrasive and exposing fresh sharp abrasive.

There are several different types of wheel dressers available (Fig. 4-6). I prefer the mechanical or star wheel type of dresser. These dressers have several star wheels made of hardened high-carbon steel. The wheels are mounted side by side on a common arbor at the end of a handle. Place the dresser on the tool rest and then, with firm pressure, move it back and forth evenly over the face of the wheel. As the wheel rotates, it causes the star wheel to rotate. This action causes the outer sur-

face of the wheel to fracture away. (When dressing a wheel, try to avoid running the tool off the face of the wheel as this will round the corner of the wheel.)

Carbide dressers should be used for dressing wheels used for sharpening carbide tools. These dressers consist of a rod of extremely hard carbide encased in a soft steel handle. They are used in the same way that mechanical dressers are used.

Important: whenever using a grinding wheel and especially whenever dressing a grinding wheel, *always wear safety glasses*. Also, when starting a grinder, *Do not stand in line* with the wheel until after it has come up to full speed.

Wheels should always be mounted with the proper fiber or cardboard disks between them and

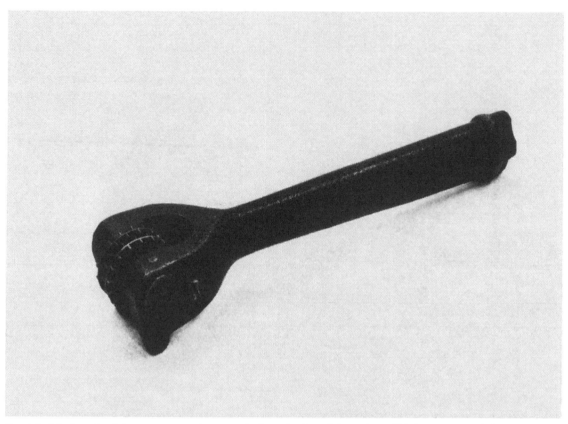

Fig. 4-6. Grinding wheel dressers are used to true grinding wheels and dress them. This is a star wheel dresser.

the clamping washers. The clamping nut should be made snug but not tight. Too much clamping pressure can crack the wheel.

Selecting Your Grinding Wheel

Grinding wheels are made with many combinations of materials that have very different properties. I think it's a good idea to have some understanding of what these variations are, but for the average home machinist, wheel selection should be quite simple. The three really important variables are abrasive material, grain sizes, and wheel hardness. The type of material you are grinding and the type of grinding you are doing—sharpening, polishing, or metal removal—will determine the type of wheel you should use.

As previously stated, the most frequent type of grinding done in the home workshop is tool sharpening. Two wheels, one for steel tools and one for carbide tools, will suffice for most. This means that you will want one aluminum oxide wheel, with fine grit (60 to 80 grit size), vitrified bonding, and graded as having medium (K) hardness. This wheel will work well for sharpening all tool steels including high speed steel. The second wheel should be a fine (80 grit size) silicon carbide wheel with a vitrified bonding and graded as soft (S). This wheel is best suited for sharpening carbide tools.

All grinding wheels, with the exception of diamond wheels, are identified with a standard marking system. This system was established by the Grinding Wheel Manufacturers Association and later adopted by the American Standards Association. By learning how to read these markings, you can determine almost anything you might want to know about a wheel. Table 4-1 illustrates how the markings read.

Type of Abrasive

For most work, aluminum oxide is the preferred abrasive. It works well on most types of ferrous materials including low strength steels, carbon steels, alloy steels, soft or hardened tool steels, cast alloy tool steels, and wrought iron. It always works well on cast and hardened bronze. Various manufacturers identify aluminum oxide by their own trade names. Some of the better known names are Aloxite, Borolon, Lionite, and Alundum.

Silicone carbide is a harder abrasive than aluminum oxide. Wheels made of silicone carbide are used to grind very hard materials such as cast iron, some cast bronzes, carbide tools, glass, and ceramic. Strange as it may seem, silicone carbide is also the preferred abrasive when grinding some of the softer materials such as copper, brass, and aluminum. Like aluminum oxide, silicone carbide is identified by a number of different trade names. Some of them are Carborundum, Carbolon, and Crystolon.

The type of material to be ground and the desired finish determine the best grit size to use. Coarse wheels are better suited for rapid material removal on soft metals. Fine wheels are better suited for rapid metal removal on hard metals. Fine wheels are also better suited for finish work.

Grade

The *grade* or *hardness* of a wheel is determined by the type and amount of bonding material used in manufacturing. Hard grade wheels are used for grinding soft metals such as iron or mild steel. The bonding material does not fracture away very rapidly, and therefore these wheels are considered long wearing wheels and well suited for general purpose work.

Soft grade wheels are better suited for grinding hard materials. The soft bonding material fractures away faster, releasing the old dull abrasive. Soft grade wheels are generally used for sharpening tools. The grade of a wheel is specified by different letters of the alphabet. "A" indicates the softest grade, and "Z" indicates the hardest grade.

Structure

The *structure* of a wheel is the density or spacing of the grit in the wheel. Structure is specified by numbers ranging from 1 to 15. The number 1 indicates the densest type of wheel, and the number 15 indicates the most porous or open grained type of wheel. Porous wheels are best suited for rapid metal removal. Dense wheels are best suited for polishing.

Type of Bonding

The *bonding material* is the material used to hold the abrasive grit together. The most common bonding is a vitrified bonding. The material is actually a special type of clay that, when heated to high temperatures, turns into a glasslike material. When this happens, the clay is said to have *vitrified*. In manufacturing a vitrified wheel, the abrasive is mixed with the clay, pressed into shape, and heated until the clay vitrifies. The resulting wheels are strong, somewhat porous, and impervious to oils, water, and most solvents. In use, the vitrified clay slowly breaks down to release dull abrasive and expose sharp abrasive. Vitrified wheels are considered good general purpose wheels.

Semivitrified wheels are similar to vitrified wheels. The clay contains a percentage of sodium silicate. The resulting bond breaks down faster than a vitrified bond exposing fresh, sharp abrasive at a faster rate. As a result the cutting action of the wheel is faster, but the life of the wheel is much shorter. Semivitrified wheels are generally used for sharpening tools such as drills, lathe tools, and milling cutters.

Rubber is frequently used to bond cut-off

Table 4-1. Standard Grinding Wheel Marking System.

Manufacturers' Prefix	Kind of Abrasive	Grain Size	Grade Hardness	Structure	Type of Bond	Manufacturers' Record
XX	A (Aluminum oxide)	Course 10 12 14 15 20 24	A (soft) B C D E F G	1 (dense) 2 3 4 5 6 7	V (vitrified) S (silicate) R (rubber) B (resinoid) E (shellac) O (oxychloride)	(Mfg's private code)
	C (Silicone carbide)	Medium 30 36 46 54 60	H I J K L M	8 9 10 11 12 13		
		Fine 70 80 90 100 120 150 180	N O P Q R S T U	14 15 (open)		
		Very Fine 220 240 280 320 400 500 600	V W X Y Z (hard)			

Most grinding wheels are identified by this standard marking system established by the Grinding Wheel Manufacturers Association. Knowing how to read these markings will enable you to determine anything you might need to know about a wheel.

wheels. The rubber provides a resilient, shock-proof bonding, strong enough to be used on high speed machines. Small rubber-bonded wheels are also used in finishing, deburring, and polishing work. They last a long time and are capable of producing a fine finish.

Another material used for high speed cut-off wheels is *resinoid*. Resinoid wheels are quite strong and shock-resistant. They are used for rough grinding operations where rapid metal removal is desired.

Shellac is used as a bonding agent when very fine finish grinding is done. The shellac is somewhat resilient in nature and slow wearing. These wheels are frequently used in centerless grinding machines.

DRILLING MACHINES

Drill motors and drill presses (Fig. 4-8) are probably the most important power tools in the shop. Drill motors have the advantage of being inexpensive and portable. For machine shop work, the drill press is much preferred. Accuracy may not be important all the time, but there are those times when it is, and it's hard to drill any kind of a precision hole with a hand drill.

Drill presses come in a variety of sizes, from small bench top models to large floor-mounted units. On some of the smaller units the power head is rigidly mounted, and the work table is mounted on a rack and pinion. The table actually pushes the work up to the drill. On large presses, the drill is usually pushed down into the work. Drill presses are used for most hole making operations such as drilling, reaming, counterboring, and countersinking. With special attachments, they can also be used for tapping operations.

Drilling

Drilling is the most common hole making operation. A reasonably sharp drill will make a hole that is accurate in diameter to within .006 inch. The finish inside the hole may be somewhat rough, but adequate for clearance holes or for tapping.

High speed drills are readily available in hardware stores and machine tool supply houses. They will drill most metals, with the exception of cast iron and hardened steels. When drilling these materials, it is preferable to use a carbide drill. Most drilling operations go better when cutting fluid is used. The cutting fluid serves several purposes. It acts as a lubricant, thereby reducing tool wear, and as a coolant, preventing overheating of the cutting edge. Cutting fluid also prevents softer materials such as aluminum from fusing to the surface of the drill. There are a number of good cutting fluids on the market, but I have found that a mixture of kerosene and transmission fluid works quite well on almost all metals.

The *turning speed* and *feed rate* of the drill are dependent on the hole size and the type of material being drilled. I have better luck using slower turning speeds and a fairly good feed rate, frequently removing the drill from the hole to clear out the chips and allow the heat to dissipate. The feed rate should be sufficient to cause a continuous chip. It should always be slowed or reduced when the drill is near the breakthrough point.

Tables 4-2 and 4-3 can be used to determine the optimum turning speed for drilling most materials. Their use may be purely academic because many drill presses do not indicate r.p.m. You can use the tables as a guide while trying to approximate the correct speed. Table 4-2 indicates the optimum speed, in feet per minute, that the cutting edge of the drill should travel when cutting a specific kind of material. Table 4-3 translates feet per minute into r.p.m. for various size drills. To use the tables, look up the desired cutting speed in feet per minute in Table 4-2. Then convert to r.p.m. in Table 4-3.

As an example, to find the best speed for drilling a ½-inch diameter hole in aluminum, Table 4-2 tells you that a cutting speed of 200 to 300 feet is desired. Table 4-3 tells you that a ½-inch diameter drill turning 764 r.p.m. has a cutting edge speed of 100 feet per minute. To get a speed of 200 feet per minute, simply double the r.p.m., for a speed of 300

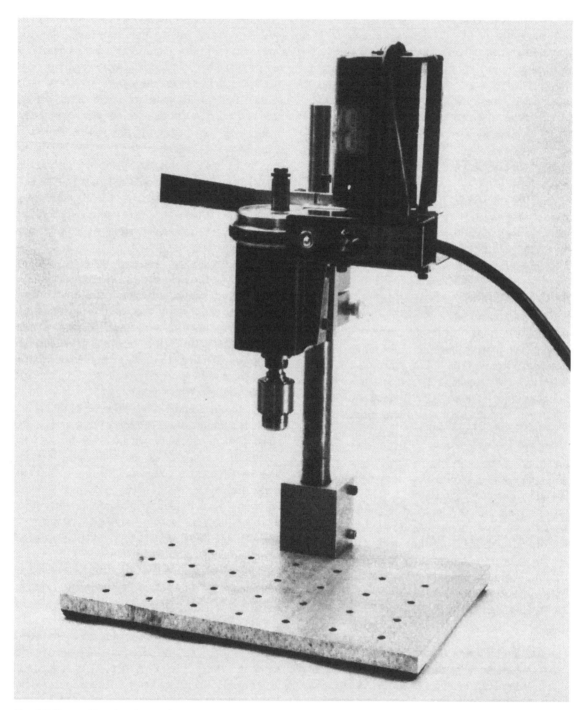

Fig. 4-7. No machine shop is complete without a drill press. The one shown here is homemade using a Sherline milling head.

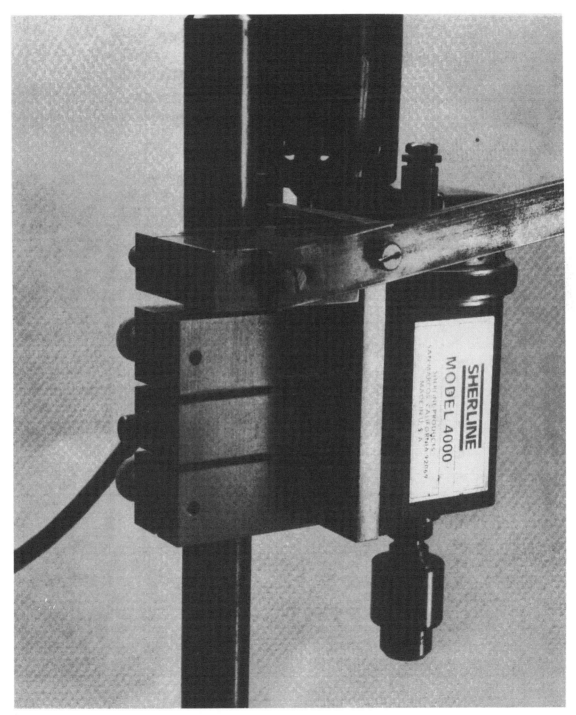

Fig. 4-8. A closeup view of my homemade drill press.

Table 4-2. Cutting Speeds for Drills.

Material	Feet Per Minute
Aluminum Alloys	200 - 300
Brass	150 - 300
Bronze	30 - 150
Cast Iron	50 - 125
Steel, low carbon	80 - 110
Steel, high carbon	70 - 80
Tool Steel	50 - 60
Stainless Steel, 300 Series	30 - 80
Stainless Steel, all others	15- 50
Wood	300 - 400

feet per minute, triple the r.p.m. Therefore, the optimum speed of the drill should be between 1528 and 2292 r.p.m.

When using a drill press, small items should always be held in a vise. Large items should be clamped to the table. The mark from a center punch will usually be sufficient to start a drill and prevent it from walking. If the hole has to be located precisely, it is best to use a center drill (Fig. 4-9) to start it. The heavy shank holds it right on center. Whenever you drill a large diameter hole, drill a small diameter pilot hole first (Fig. 4-10). Large drills usually have a broad chisel edge at the point.

The chisel edge makes it hard for the drill to penetrate the material being drilled.

Countersinking and Counterboring

Countersinking and *counterboring* are methods of modifying a hole to insure an accurate seat for the head of a screw or bolt, or to permit recessing the head of a screw or bolt below the surface of the part.

A *countersink* (Fig. 4-11) is a pointed cutting tool that cuts a beveled edge around the top of a hole to match the bevel on the underside of a flathead screw. Most flathead screws have an 82° included angle. Others may have a 100° included angle. Countersinks are available to match both. Still other countersinks with a 60° included angle are available for drilling center holes in stock that is to be turned between centers on a lathe. For good workmanship, always take care to select the right countersink for the job.

Countersinks are self-centering tools, so the only thing the machinist has to watch is the depth of the cut. Cutter speed should be about one-half the recommended drilling speed or slower. If the tool is turning too fast, it will have a tendency to chatter. The resulting countersink will be uneven.

Table 4-3. Conversion Table F.P.M. to R.P.M.

Dia. Inches	10'	20'	30'	40'	50'	60'	70'	80'	90'	100'	Feet per Minute
1/64	2445	4889	7334	9778	12223	14668	17112	19557	22001	24446	Revolutions
1/32	1222	2445	3667	4889	6112	7334	8556	9778	11001	12223	per
1/16	611	1222	1833	2445	3056	3667	4278	4889	5500	6112	Minute
3/32	407	815	1222	1630	2037	2445	2852	3254	3667	4074	
1/8	306	611	917	1222	1528	1833	2139	2445	2750	3056	
5/32	244	489	733	978	1222	1467	1711	1956	2200	2445	
3/16	204	407	611	815	1019	1222	1426	1630	1833	2037	
7/32	175	349	524	698	873	1048	1222	1397	1572	1746	
1/4	153	306	458	611	764	917	1070	1222	1357	1528	
5/16	122	244	367	489	611	733	856	978	1100	1222	
3/8	102	204	306	407	509	611	713	815	917	1019	
7/16	87	175	262	349	437	524	611	698	786	873	
1/2	76	153	229	306	382	458	535	611	688	764	
5/8	61	122	183	244	306	367	428	489	550	611	
3/4	51	102	153	204	255	306	357	407	458	509	
7/8	44	87	131	175	218	262	306	349	393	437	
1	38	76	115	153	191	229	267	306	344	382	

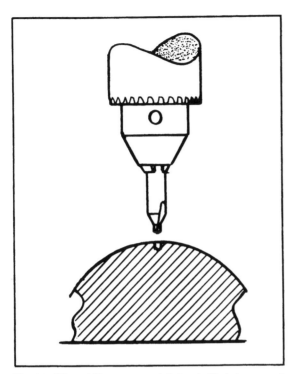

Fig. 4-9. When drilling an accurately located hole in any part, start the hole using a center drill. The large diameter shank will prevent the drill from walking or slipping sideways and ensure that the hole is located where you want it.

Counterbores (Fig. 4-11) are used to recess sockethead cap screws or to provide a flat surface under the head of a screw on an irregularly shaped surface. Unlike the countersink, the counterboring tool is not self-centering. It requires a pilot to hold it on center. Some counterboring tools are made with a built-in pilot, while others have removable pilots. Those with removable pilots usually come with a set of different sized pilots so you can select the proper size for the hole on which you are working. A pilot 0.002 inch to 0.006 inch smaller in diameter than the drilled hole is about right. The cutter speed should be approximately one-third the speed recommended for a comparable sized drill. Be sure to use a generous amount of cutting fluid.

Reaming

Reaming is a method of producing a close tol-

eranced hole with a good finish. As stated before, a drilled hole may have as much as .002-inch to .008-inch variation in diameter. When it is necessary to produce a hole that will be a slip fit or a press fit for a dowel pin, bushing, or bearing, you must have a way of holding closer tolerances on the diameter of the hole. With a reamer (Figs. 4-12 and 4-13) you should be able to produce a hole within two or three ten-thousandths (0.0002 inch or 0.0003 inch) of the desired diameter. Reamers are available in all fractional sizes starting with a diameter of 3/64 inch. They are also available in standard wire gauge sizes and drill sizes. They are generally made of high speed tool steel. They are also available in carbide. Some reamers are intended for use in a hand-held handle similar to a tap handle, and some are intended for chucking in a machine.

When using a machine reamer, the hole should be drilled 0.010 inch to 0.015-inch smaller than the desired finished hole size. Then, using a slow r.p.m. and a slow feed rate, the reamer can be run into the hole. Usually the slower the feed rate the better the

Fig. 4-10. Drilling a small diameter pilot hole will make drilling a large diameter hole easier.

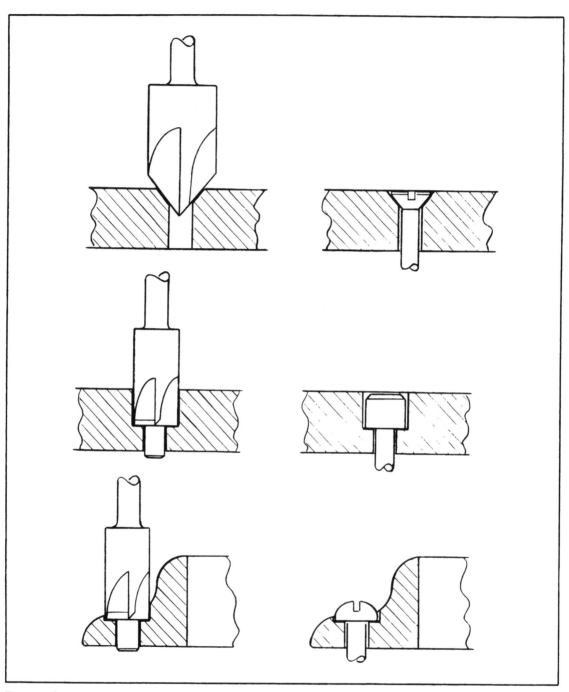

Fig. 4-11. Countersinks are used for removing burrs around the edge of a drilled hole and to provide a means of recessing flat-head screws. For best results, make sure the bevel on the countersink matches the bevel on the screw. Counterbores are used to recess machine screw heads. They are also used to provide a flat surface for screw heads.

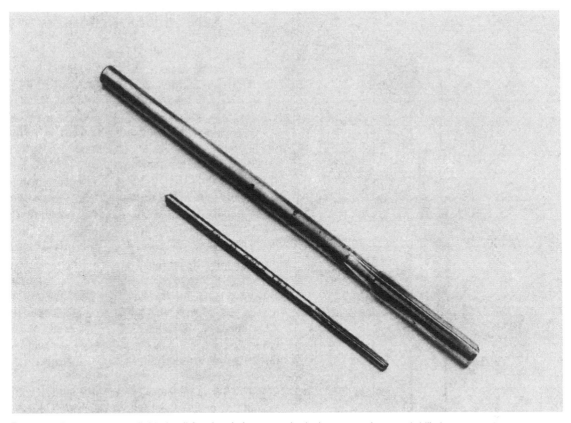

Fig. 4-12. Reamers are available in all fractional sizes, standard wire gauge sizes, and drill sizes.

finish, but if the tool is fed too slow it will dull. For most materials a feed rate of 0.002 inch to 0.004 inch per flute per revolution is about right. Use a generous amount of cutting fluid.

If the resulting hole is oversize, it is quite probable that the cause is misalignment of the tool or excessive run out in the drill press. A poor finish in the hole is generally the result of feeding the reamer too fast.

TAPS AND DIES

Taps and dies are tools used for cutting threads. *Taps* are used for cutting internal threads like those found in a nut (Fig. 4-14). *Dies* are used to cut external threads like those found on a bolt or a machine screw (Fig. 4-15). Let's review threads in general.

Most machine screws used in the United States conform to the Unified Thread System. Under this system, threads are specified by the outside diameter of the thread and then, following a dash, the pitch. In this system the *pitch* is the number of threads per inch. As an example, a ⅜-16 screw is one with a ⅜-inch major diameter having 16 threads per inch (Fig. 4-16).

Bolts and screws with thread diameters smaller than ¼ inch are specified by machine screw gauge numbers. The numbers start with 000 as the smallest and go up to 12 as the largest. The gauge number is determined mathematically, based on the number of times .013 inch will go into the difference between .060 inch and the actual screw diameter. Remember that a size 0 screw has a diameter of .060 inch. Each higher number is .013 inch larger.

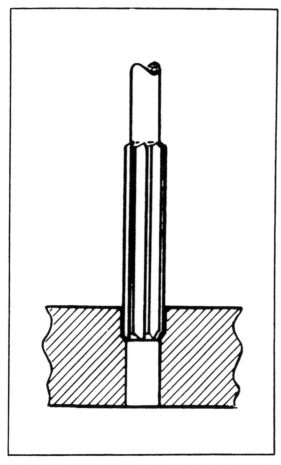

Fig. 4-13. Reamers are used to machine holes to accurate diameters and ensure a good surface finish inside the hole.

Each size smaller is .013 inch smaller. See Table 4-4.

Under the Unified Thread System there are two series of threads for most sizes. One is the fine thread series and the other is the coarse thread series. Usually when threads are called out on a drawing, the initials "UNF" or "UNC" follow the thread callout. They stand for Unified National Fine or Unified National Coarse. The proper callout for a .190-inch diameter machine screw with a fine thread is "10-32 UNF." A number 10 screw with a coarse thread is a "10-24 UNC." The thread chart in the Appendix lists all of the standard threads starting with 0 and going up to ¾ inch. The same chart lists all of the recommended tap drill and clearance drill sizes for these screws.

Most threaded parts—nuts, bolts, and screws—have right-hand threads. That is, the screw moves away from you when turned clockwise. Some special parts do have left-hand threads. When reading a blueprint, you can always assume that the thread is right-handed unless it is specifically called out as left-handed.

The use of metric screws is still not too common in this country, but a brief discussion of them seems in order. All metric threads are called out by first specifying the major diameter in millimeters, then the pitch. The pitch of a metric thread is the distance, measured in millimeters, between similar points on two adjacent threads. The letter M before

Fig. 4-14. Taps are used for cutting standard size internal threads.

Fig. 4-15. Dies are used for cutting external threads. They are available for all standard threads.

Table 4-4. Screw Sizes and Diameters.

Size		Diameter (in inches)
000	.060 − (2 × .013)	= .034
00	.060 − .013	= .047
0	.060	= .060
1	.060 + .013	= .073
2	.060 + (2 × .013)	= .086
3	.060 + (3 × .013)	= .099
4	.060 + (4 × .013)	= .112
6	.060 + (6 × .013)	= .139
8	.060 + (8 × .013)	= .164
10	.060 + (10 × .013)	= .190
12	.060 + (12 × .013)	= .216

the diameter indicates that the thread is in accordance with the International Organization for Standardization. The letter P placed between the diameter and pitch indicates the threads are in accordance with the Industrial Fastener Institute standards (Fig. 4-18).

There are three types of taps—the taper tap, the plug tap, and the bottoming tap (Fig. 4-19). The *taper tap* is the one most commonly used. The first 8 to 10 threads are tapered undersize to make it easier to align and start the tap. These taps are used wherever you wish to thread a hole completely through a part. They are also used as the first step prior to using a bottoming tap.

Plug taps have the first three to five threads tapered. Although they are a little harder to align and require more force to turn, they can also be used to thread through holes. They are really in-

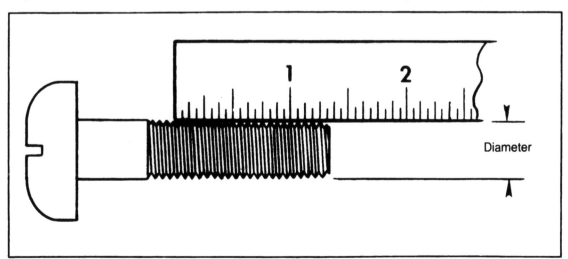

Fig. 4-16. To determine the size of a unified thread system screw thread, measure the outside diameter of the thread in inches. To determine the pitch, count the number of threads in 1 inch.

71

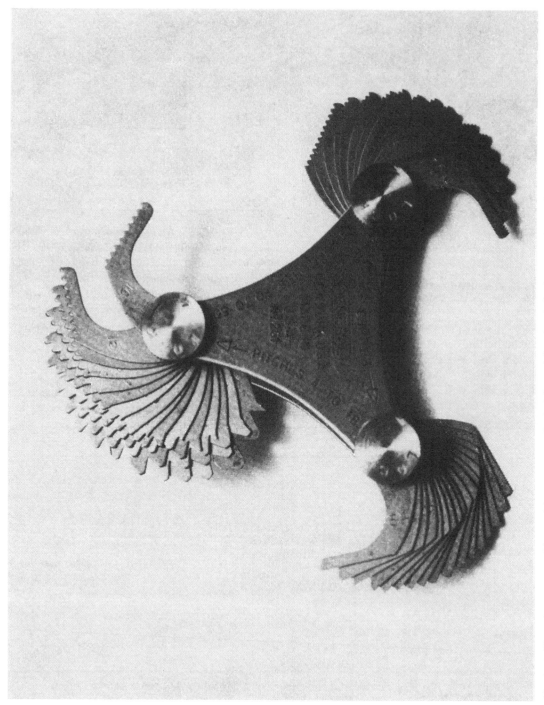

Fig. 4-17. Thread gauges provide a quick, accurate method of determining the pitch of a screw thread. The gauges, consisting of a number of templates, are held against the unknown thread, one at a time, until one of them matches.

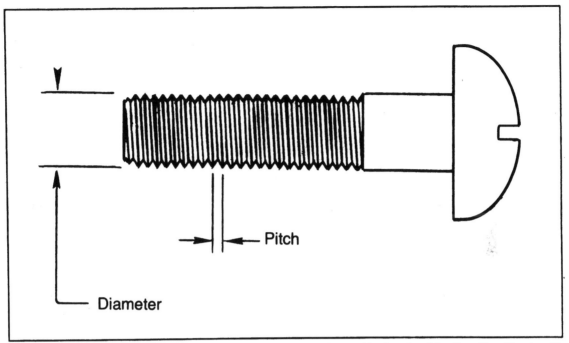

Fig. 4-18. To determine the size of a metric screw thread, measure the outside diameter of the thread in millimeters. To determine the pitch, measure the distance in millimeters between similar points on any two adjacent threads.

tended for use in threading blind holes where the depth of the thread is not critical.

Bottoming taps have only the first 1 to 1½ threads tapered. These taps are used only when it is necessary to run a thread all the way to the bottom of a blind hole. Whenever it is necessary to use a bottoming tap, you should always use a taper tap or a plug tap to prethread the hole. It takes a little longer but if you can prevent breaking a tap in the part, it is well worth the extra effort. Every machinist soon learns that taps are extremely brittle and easily broken, and that broken taps can be very difficult to remove.

A tap should be firmly secured in a tap handle (Fig. 4-20). The two most common types of handles are the T handle and the straight handle. When working with small taps, the T handle is probably the best choice. The extra length makes it easier to align, and there is less chance of the tap breaking.

Before tapping a hole, make sure the hole has been drilled out to the proper size for the tap you

are using (Fig. 4-21). The proper tap drill size can be found in the thread chart in the Appendix. Whenever practical, cut a small chamfer around the opening of the hole. This can be done with a standard countersink tool. Then the tap, mounted in a tap wrench, should be liberally lubricated with a suitable cutting fluid. The tap is started into the hole with a combination of downward force and turning motion (Fig. 4-22).

After the first two or three complete revolutions, the tap should be backed out a quarter turn, and after that it should be backed out a quarter turn for every complete turn in. This causes the chips to break and helps to prevent the tap from jamming in the hole. When using a bottoming tap, remove the tap and clear out the chips after tapping about three-fourths of the total depth of the hole (Fig. 4-23).

Always use extreme care to avoid side pressure on the tap. Should the tap start to turn hard, *don't force it*. Loosen it by gently rotating it back and

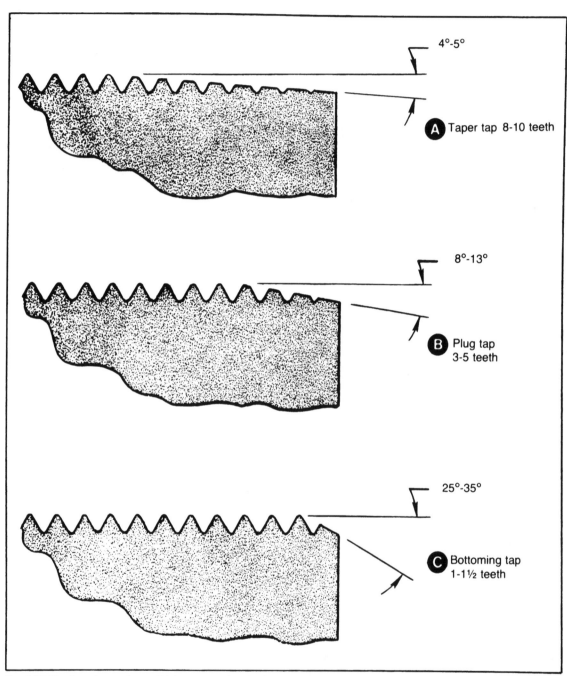

Fig. 4-19. The three types of taps. (A) the taper tap has the first 8 to 10 threads tapered undersize to make it easier to align and start. (B) The plug tap has the first three to five threads tapered undersize. The tap can be used to start a thread and cut a usable thread deeper into the hole. (C) Bottoming taps have only the first 1 to 1½ threads tapered. These taps are used to cut a full thread to the bottom of a blind hole.

74

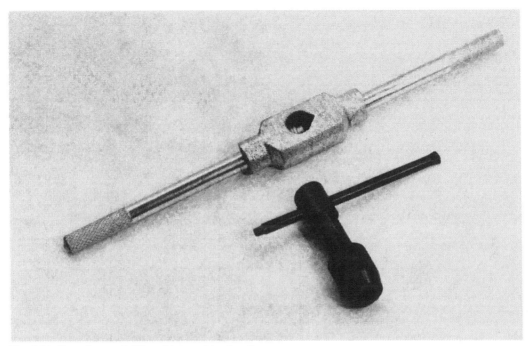

Fig. 4-20. Two of several different types of tap handles.

forth. Then remove it and clear the chips before starting again.

Dies are used to cut external threads like those on the outside of a bolt. There are several types of dies available. Most of them are used with a special handle called a *die stock* (Fig. 4-24). Some dies are made adjustable so that you can control how tightly the part being threaded will fit into its mating part. All dies are made with the first few threads chamfered or tapered to make it easier to start the thread.

The actual cutting of threads with a die is much the same as cutting threads with a tap (Fig. 4-25). The part to be threaded should be held securely in a vise. Then the area to be threaded should be well lubricated with a good cutting fluid. The die, mounted in a die stock, is started on the part. Always start the die with the chamfered threads (Fig. 4-26). Use care to keep the die aligned straight with the axis of the part. After the first two or three turns of the die, back it up approximately a quarter turn to break the chips. Then the die should be backed a

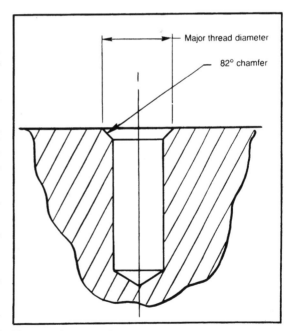

Fig. 4-21. Cut a chamfer around a hole prior to tapping it. The chamfer should open the mouth of the hole to at least the major diameter of the desired thread.

75

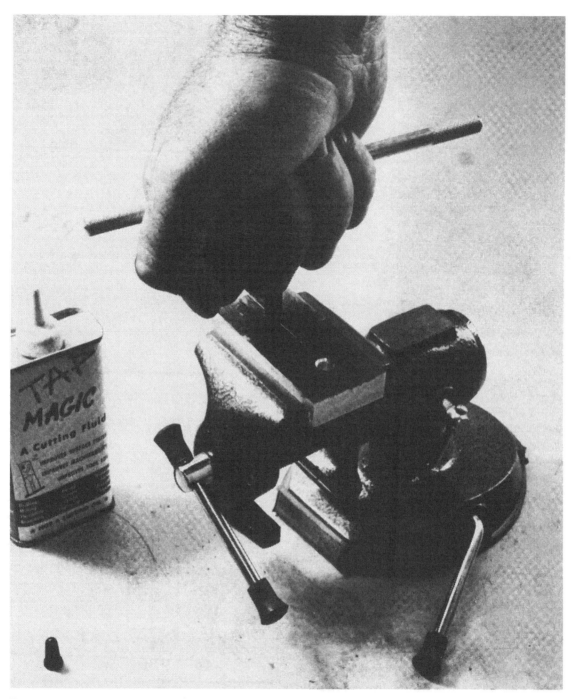

Fig. 4-22. To start a tap, hold it in alignment with the hole, then apply downward pressure as you turn it. Once the tap is started, it is self-feeding and the downward force is no longer required.

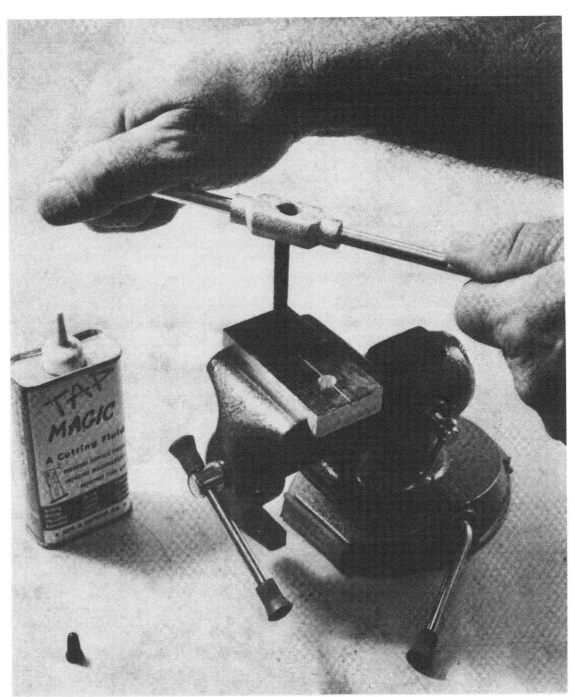

Fig. 4-23. Once the tap is started, it should be turned with both hands to prevent side pressure, which might break it. Back the tap out a quarter turn for every full turn in or whenever heavy resistance is felt. Reversing the tap will cause the chips to break and prevent the tap from jamming in the hole. Always use liberal amounts of tapping fluid.

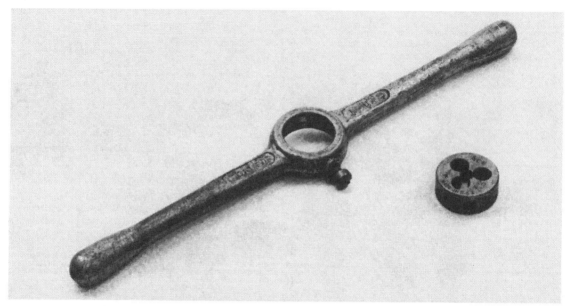

Fig. 4-24. Die handles are made in several sizes, which may vary from one die manufacturer to another.

Fig. 4-25. All the same considerations for using a tap also apply when using a die: proper alignment, downward force for starting, and periodic reversal to break the chips.

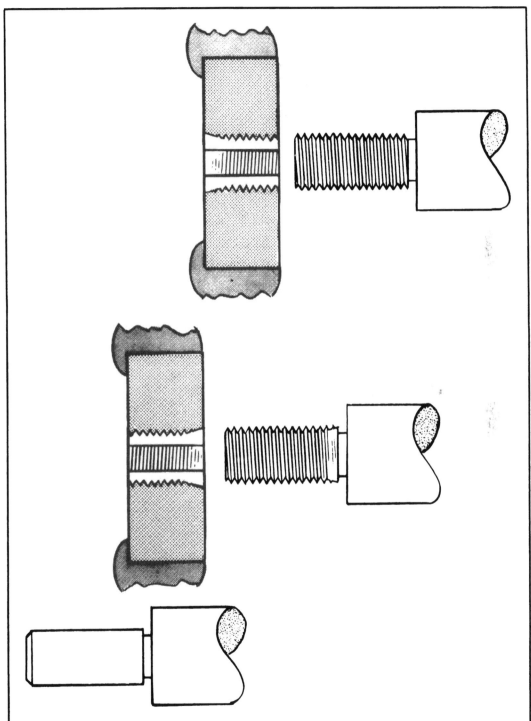

Fig. 4-26. A die should always be started from its chamfered side. If it is necessary to cut a full thread down to a shoulder, the die can be reversed after initially cutting the thread, chamfered side first.

quarter turn for every full turn in until the part is fully threaded.

If it is necessary to have a full thread all the way up to a shouldered area on a part, first use the die as described, then remove it, reverse it, and recut the threads with the back of the die. Cut a relief into the part next to the shoulder. The relief provides a good way to terminate the thread.

Chapter 5

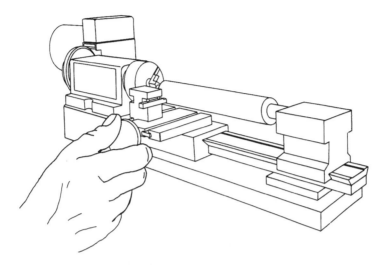

The Lathe

THE INVENTION OF THE LATHE WAS PROBABLY one of the most important highlights in the history of machine tools. The lathe made possible the accurate machining of cylindrical objects, and there are few modern machines that do not contain at least one or two parts that were made on a lathe.

The first lathes were crude machines called *turns*. To use one of these early lathes, the machinist had to use one hand to turn the machine while he cut the part with a *graver* held in the other hand. Modern lathes bear little resemblance to these early machines. Today there are power-driven lathes with micrometer feedscrews to position the cutter. You can turn out precision parts with no trouble at all.

Modern lathes are made in varying degrees of sophistication and a wide range of sizes. As a rule, the lathes found in the home workshop are small ones, designed for versatility. The lathe that I have selected for use in illustrating this book is a Sherline, marketed by several large retailers such as Sears, Jensen Tool, and Brookstone under their own trade names. Small, inexpensive, yet versatile enough to do a wide variety of jobs, I consider it to be an ideal machine for the home workshop.

Figure 5-1 shows the major parts of a modern lathe. Basically it consists of a bed, headstock, tailstock, saddle, and a cross-slide. The *bed* provides a base for the entire machine and maintains the proper alignment between its major components. The *headstock* provides a means of holding and turning your part. The *tailstock* is used to support the free end of a long workpiece and also provides a means of drilling your workpiece.

The *saddle* or *carriage* is driven by a feedscrew and travels the length of the bed. It provides a means of positioning the cutting tool along the axis of the lathe. Mounted on the saddle is the *cross-slide*. It is also controlled by a feedscrew, and it provides a means of positioning the cutting tool crossways to the axis of the lathe.

Larger lathes also have a compound slide that can be set at various angles to the axis of the lathe. The compound slide is used for cutting tapers and

Fig. 5-1. The major parts of a lathe are: (A) motor, (B) headstock, (C) spindle, (D) tool holder, (E) cross-slide, (F) saddle, (G) bed, (H) tailstock ram, and (I) tailstock.

bevels. On the Sherline lathe this function is accomplished by rotating the headstock in relation to the bed of the machine. This operation is discussed in detail in Chapter 5.

SELECTING A LATHE

One of the major considerations in selecting a lathe is size. When you purchase a lathe, you want to know that it has sufficient capacity to handle the size objects that you intend to make. The four most important dimensions to consider are the swing over the bed, the swing over the saddle, the distance between centers, and the diameter of the hole in the spindle (Fig. 5-2). These dimensions limit the size of the parts that you can make on your lathe. With some lathes like the Sherline, the turning capacity (swing over the bed and swing over the saddle) can be increased with the use of an inexpensive accessory (Fig. 5-3). Because the tailstock height is not increased, this accessory works for faceplate-, chuck- or collet-mounted parts only.

The temptation to purchase a lathe larger than you actually need should be resisted for several reasons. First, the larger the machine the higher the cost—not only the initial cost, but the cost of any accessories that you may wish to add at a later date. Second, I have found it much easier to do small work on a small lathe. The probability of damaging your work by taking too large a cut is greater on a large lathe. Few things are more discouraging than to ruin a part by taking too deep a cut just before finishing it. With a small lathe, the tendency is to take smaller cuts more consistent with the size of the part you are making.

The last reason for buying the smaller lathe is storage space. The larger the machine, the more bench space or floor space it requires. Small machines do not require a permanent setup. They can easily be stored when not in use and brought out only when you need them.

Fig. 5-2. The size of a lathe is determined by the dimensions of: (A) swing over the bed, (B) swing over the carriage, (C) distance between centers, and (D) diameter of hole through spindle.

Fig. 5-3. The swing over the bed and saddle can be increased on some lathes, like this Sherline, with the use of a simple spacer block.

Fig. 5-4. Adjustable gibs are essential on a lathe to provide a means of compensating for wear. Both the saddle gib and the cross-slide gib on a Sherline lathe are shown.

When you have determined the size of lathe you want, and you actually start shopping for one, you'll want to pay close attention to features like construction and versatility.

The backbone of any lathe is its bed. A good lathe has to have a rigid bed. Otherwise, it will be virtually impossible to maintain the proper alignment between the headstock, the tailstock, and the

carriage. A weak bed will flex so that you can't hold close tolerances. It will also permit vibration and chatter, making it difficult to obtain a good finish on your part.

The lathe should also have adjustable gibs (Fig. 5-4), so that you can remove any play between the carriage and the bed, or between the cross-slide and the carriage. Any looseness in these areas will

Fig. 5-5. Some lathes, like the Sherline, may be converted for use as a milling machine.

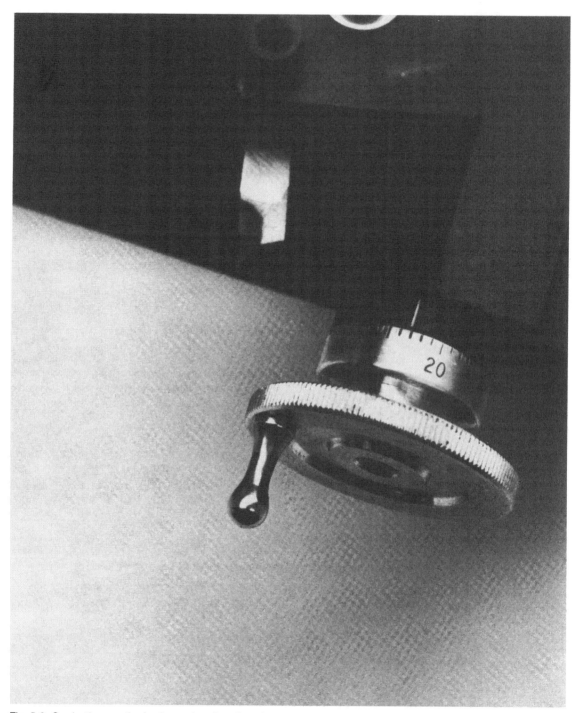

Fig. 5-6. Graduations on the feedscrew handles give micrometer accuracy in positioning the cutting tool.

be a source of vibration leading to chatter and making it difficult to hold close tolerances.

The headstock is the second most important part of the lathe. There should be provision in the headstock to permit tightening the spindle bearings. This may not seem important on a new lathe, but as the machine gets older you will want to be able to compensate for wear.

Still another consideration is the feedscrews. A machine that has exposed feedscrews is more vulnerable to wear than one that has them enclosed.

Most of the modern lathes, especially small lathes intended for use in the home, are designed to be quite versatile. Some lathes, like the Sherline, have attachments that even permit them to be used as a horizontal or vertical milling machine (Fig. 5-5). As a vertical milling machine, it can also function as a precision drill press. These functions are discussed in more detail in Chapter 7.

LATHE CUTTING TOOLS

The lathe tool, or tool bit as it is sometimes called, is the key to good lathe work. Some tool bits are made from high speed tool steel, while others have carbide cutting inserts brazed onto a steel shank (Fig. 5-7). Both kinds are readily available from most machine tool supply houses. The question of whether to use a high-speed steel or carbide tool bit is easily answered by considering the metal to be worked.

For most metals, especially the softer ones like aluminum, brass, iron, and mild steel, the high speed bit will do an excellent job. These tools retain their hardness at temperatures up to around 1000° F. They can be used to take fairly heavy cuts at speeds comparable to the speeds used for drilling. (Cutting speeds are measured in feet per minute and may be taken from Tables 4-2 and 4-3.) The big advantage of using high speed bits is that they can be shaped and sharpened easily on almost any kind of grinding wheel.

Carbide tool bits are harder and longer wearing than the high speed bits. They retain their hardness at temperatures up to approximately 1700° F.

Fig. 5-7. High speed steel lathe tool (left) and carbide-tipped lathe tools (right) are available for all sizes of lathes. The best tool to use depends on the type of material being machined.

Therefore, cutting speeds can be almost double the speeds used for high speed bits. They also retain their cutting edge longer. These tools are used for production run cutting of hard metals such as alloy steels, tool steels, and stainless steels. The drawbacks to carbide bits are that they are generally more expensive and harder to grind and sharpen. You can't grind them on an ordinary aluminum oxide grinding wheel. Instead you need a silicone carbide wheel. (Refer to the section "Selecting Your Grinding Wheel" in Chapter 4.)

Carbide tools are generally used in production shops. For home use, it is debatable whether they are worth the additional cost and trouble. My advice is to stick with the high speed tool bits unless you are doing a considerable amount of work on hard materials.

SHAPING AND SHARPENING LATHE TOOLS

No two machinists have the same ideas on shaping lathe tools. Each machinist has his own preferences and thinks the shapes he uses are best.

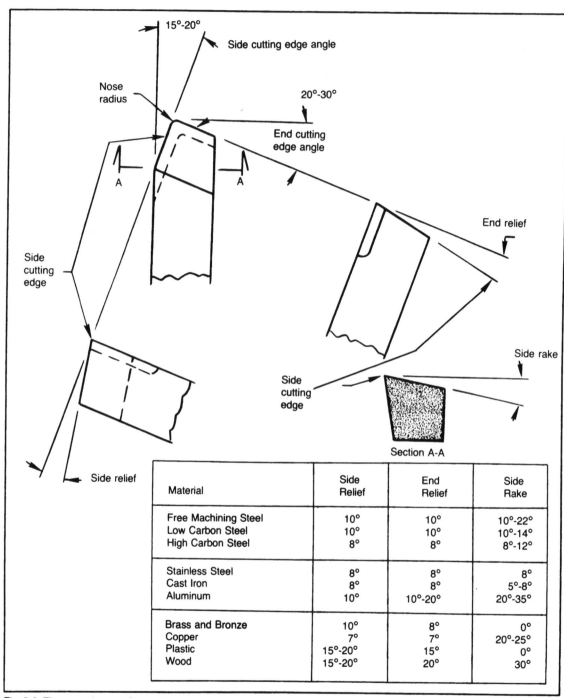

15°-20°

Side cutting edge angle

Nose radius

20°-30°

End cutting edge angle

Side cutting edge

A   A

End relief

Side rake

Side cutting edge

Section A-A

Side relief

Material	Side Relief	End Relief	Side Rake
Free Machining Steel	10°	10°	10°-22°
Low Carbon Steel	10°	10°	10°-14°
High Carbon Steel	8°	8°	8°-12°
Stainless Steel	8°	8°	8°
Cast Iron	8°	8°	5°-8°
Aluminum	10°	10°-20°	20°-35°
Brass and Bronze	10°	8°	0°
Copper	7°	7°	20°-25°
Plastic	15°-20°	15°	0°
Wood	15°-20°	20°	30°

Fig. 5-8. There are three surfaces to be ground when sharpening a lathe tool. This illustration identifies them and the resulting cutting edges. The table shows the preferred angles for each of these surfaces based on the type of material to be cut.

Actually, the shape is not all that critical. The really important thing is to keep the tool sharp and make sure that you have good clearances at the side and bottom of the bit. The following discussion is intended to point out some of the basic considerations and give you a few guidelines as you sharpen your first few tools.

Figure 5-8 illustrates and identifies the various parts of a typical tool. The tool shown is called a *right-hand turning tool* and is intended for reducing the diameter of a part by cutting from right to left or toward the headstock. The side cutting edge angle should be approximately 20°. This angle permits the side of the cutting edge to enter into the cut first and allows the loading on the nose of the tool to build up gradually. This helps to prevent chipping or breaking the tool bit. Too great an angle will change the direction of the cutting force, which may cause the part to deflect away from the tool (Fig. 5-9). This deflection would result in your cutting a taper or a nonuniform diameter on your part.

For large lathes, when the tool is intended to be used as a roughing tool, the end cutting edge angle should be kept small—approximately 10° to 20°. This leaves plenty of material to support and strengthen the nose of the tool and still provides some clearance between the end cutting edge and the part. The nose radius should be approximately 1/64 inch.

For the amateur machinist using a small lathe, most work can be done with a general purpose finishing tool. On a general purpose finishing tool, the side cutting edge angle remains the same, the end cutting edge angle is increased to about 50°, and the nose radius is increased to about 1/32 inch. The more generous the nose radius, the smoother the cut. This tool is actually ideal for both turning and facing. It's a tool that you can use to do most of your work.

Occasionally you may find yourself making a part where it will be more desirable to cut from left to right or to cut a left-hand face on a part. For this work you will need a left-hand tool. Simply grind the tool as a mirror image of a right-hand tool.

The side rake should be varied depending on the type of material you are cutting. For steel, the rake should be approximately 10°. Copper and aluminum are soft, gummy metals that tend to stick and build up on the cutting edge of the tool. To prevent this buildup, increase the side rake to about 20° and use a small amount of cutting oil brushed on the tool as you cut.

Brass and bronze cut best with no side rake. This is because side rake has a tendency to make the tool bite into the brass and result in an uneven cut. Because no side rake is required for cutting brass, you can actually use the same tool for both right- and left-hand turning and facing (Fig. 5-10).

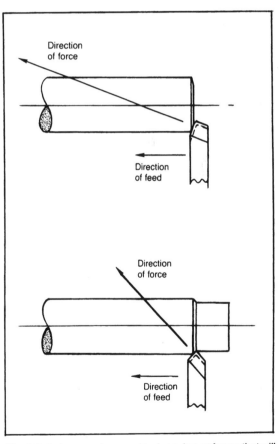

Fig. 5-9. An improperly ground tool may impart forces that will tend to deflect your work. A poor finish and dimensional inaccuracies are the result of an improperly ground tool.

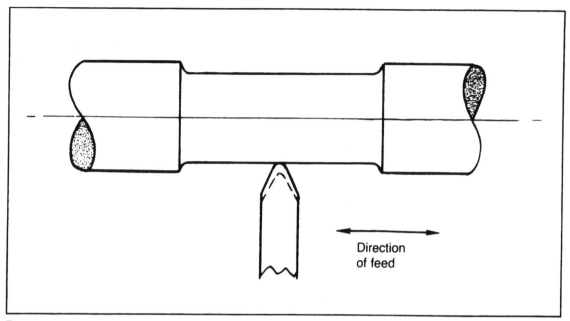

Fig. 5-10. Since tools ground for cutting brass require no side rake, they can be used for both left- and right-hand cutting.

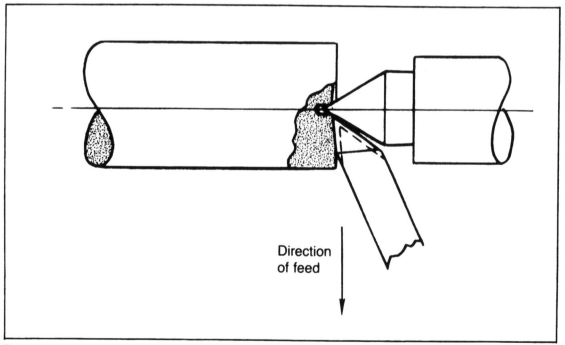

Fig. 5-11. When facing a part mounted between centers, the lathe tool should be ground to a sharp point without a nose radius. This will permit cutting as close to the center as possible.

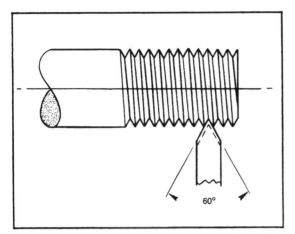

Fig. 5-12. All standard threads are cut with a 60° threading tool. For fine threads, the tool can be ground to a sharp point; for coarse threads it is desirable to radius the point slightly.

The side relief should be about 10° for cutting most materials. The only exception to this is when you are cutting wood or plastic. For these materials, both the side rake and the side relief should be increased to approximately 15°.

Whenever it is necessary to face a part that is being supported by a center (Fig. 5-11), the tool should be ground without a nose radius. This will permit the tool to come in as close to the center as possible.

Most threads are cut with a 60° threading bit (Fig. 5-12). For fine threads these bits are usually ground with a sharp point. For large threads or coarse threads it is usually desirable to radius the point slightly (Fig. 5-13).

The best parting tools (cut-off tools) are slen-

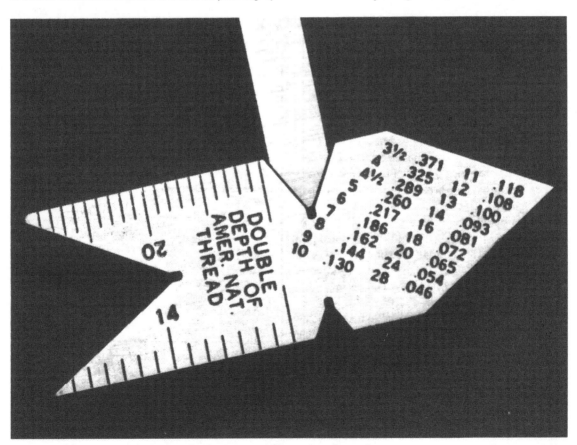

Fig. 5-13. Using a center gauge as a template when grinding a threading tool.

Fig. 5-14. A Sherline parting tool.

A soft or medium-soft aluminum oxide grinding wheel is recommended for shaping and sharpening your lathe tools (see Chapter 4). The wheel should be examined to see if it needs to be dressed or trued. When it has been determined that the wheel is in good shape, you can start grinding your tool. Start by grinding the side cutting edge and side relief. When this is completed, grind the end cutting edge. Next, grind the nose radius, blending between the side relief angle and the end relief angle. The last surface to be ground is the side rake. Hold the tool so that the wheel turns into the cutting edge of the tool. This will prevent the formation of burrs on the cutting edge.

If your grinder has an adjustable tool rest, it can be used to obtain the proper angles. If not, you can approximate the angles by hand. Take care not to overheat the tool. Overheating will soften the

der blades made specifically as cut-off tools (Fig. 5-14). These tools, used with a special tool holder, should be used as cut-off tools only. Any side loads may cause the blade to break. For small parts, up to about ⅜-inch diameter, you can grind your own cut-off tool, which will do a fair job (Fig. 5-15). Simply grind the cutting end of the tool down to a slender blade. Don't forget to put about 3° side relief on both sides. The blade should be slightly narrower than the cutting edge. The cutting edge relief should be about 10°.

Lathe bits may also be ground to make special cuts (Fig. 5-16). As an example, if your drawing calls for a certain size fillet radius or corner radius, these radii can be ground into the tool. This practice is done quite frequently in production shops. In the home shop, making one-of-a-kind projects, it is usually easier to do this work with a file. Lathe filing and its applications will be discussed in more detail in Chapter 6.

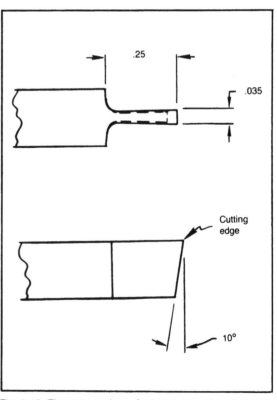

Fig. 5-15. The correct shape for home-ground parting tool.

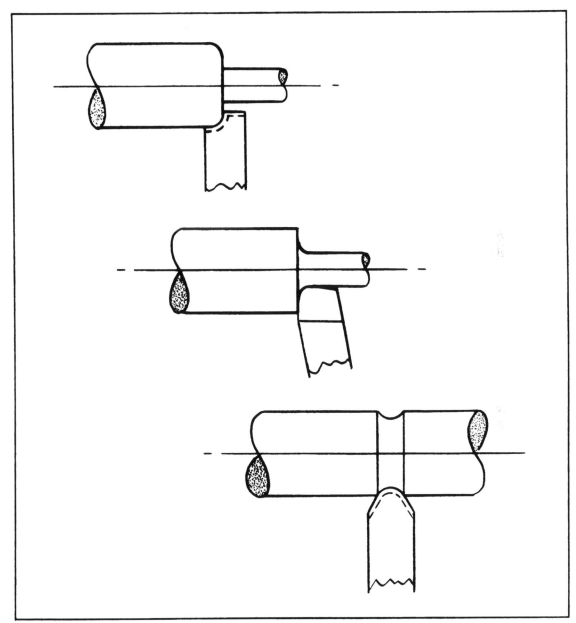

Fig. 5-16. Lathe tools are frequently ground to machine specific radii or shapes.

metal, and it will not hold a good edge. Overheating can be prevented by frequently immersing the tool in water. Any discoloration of the metal means you are getting the tool too hot. After grinding, use a fine oilstone and hone the cutting edge. This will remove grinding imperfections and burrs.

Chapter 6

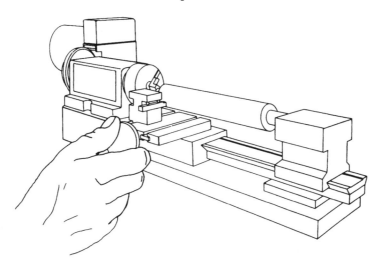

Lathe Operation

I N THE PREVIOUS CHAPTER I DISCUSSED THE IM-
portance of lathes, the basic parts of a lathe, and
lathe tools. In this chapter I will cover how to use
the lathe, how to mount your work, how to set the
tool, and how to perform various operations.

MOUNTING THE WORK

On a drill press, the work is held stationary and
the tool bit is rotated. On a lathe do just the opposite. Rotate the work and use a stationary tool
to cut it. This requires holding the part in-line with
the axis of the lathe and applying power to rotate
it. There are several good methods of ac-
complishing this. Each has its own advantages and
disadvantages, which I will discuss in some detail.

Turning between Centers

The first method of holding a part is called
turning between centers. The centers are special
arbors. One end of the center is tapered to fit a taper
in either the headstock or the tailstock. The other
end of the center is ground into the shape of a cone
with a 60° included angle. The tip of this cone is
used to support the part.

When you mount a center in the lathe, make
sure that both the tapered hole and the taper on the
center are clean (Fig. 6-1). Dirt or chips will prevent the center from seating properly and may
cause it to be off-center. To remove a center from
the headstock, slide a piece of bar stock into the
opposite end of the spindle and tap the center loose.
Make sure to hold onto the center so it doesn't fall.
Centers mounted in the tailstock are removed by
simply retracting the ram into the tailstock. This
automatically ejects anything mounted in the ram.

The part to be turned is drilled at both ends
with a center drill as close to center as possible.
The center-drilled hole should be approximately
1/16 to 1/8 inch deeper than the pilot portion of the
drill. Never drill as deep as the full diameter of the
drill. A properly drilled hole will look approximately as shown in Fig. 6-2.

The power to rotate the part is transmitted
from the lathe spindle, through the faceplate, to a

Fig. 6-1. It is important to always clean the taper before inserting a center.

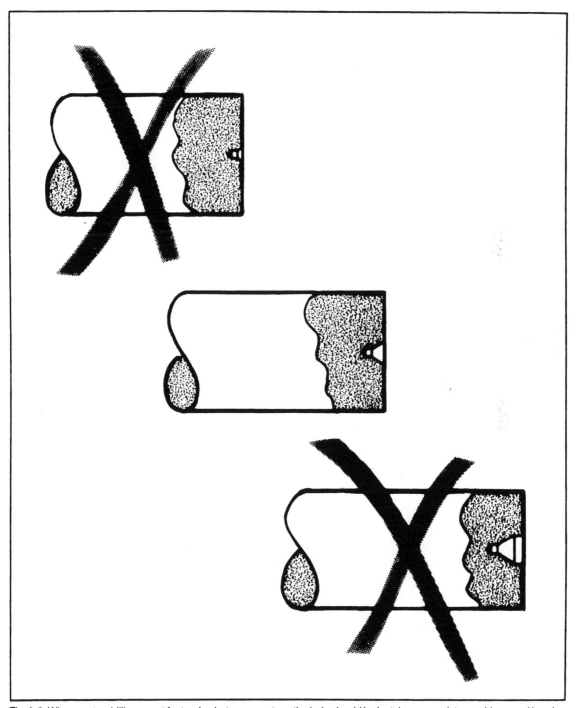

Fig. 6-2. When center drilling a part for turning between centers, the hole should be just deep enough to provide a good bearing surface for the center.

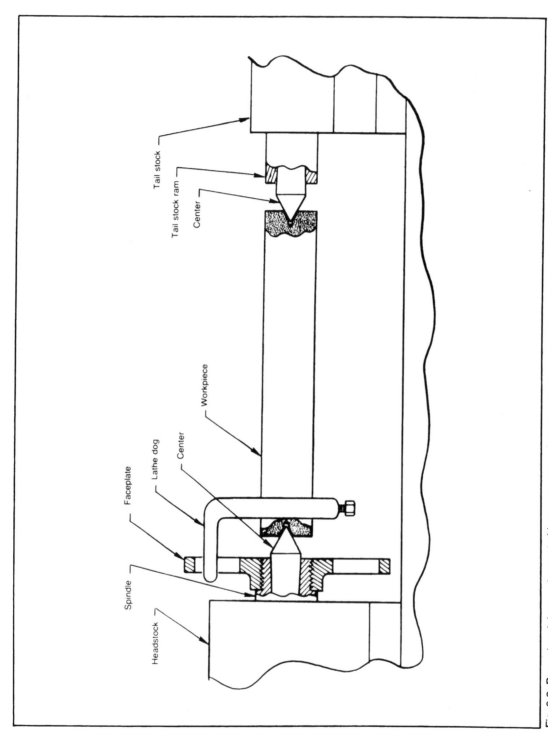

Fig. 6-3. Power to rotate a part mounted between centers is transmitted from the lathe spindle through a faceplate and a lathe dog to the part. The tailstock center should be just tight enough to eliminate any side play or movement of the part. Frequent oiling of the tailstock center is a necessity.

lathe dog, which is clamped to the part. See Fig. 6-3. Because the part is rotating against a stationary center in the tailstock, it should be lubricated frequently to prevent overheating. The headstock center rotates with the part and does not require lubrication.

If you are turning plastic parts, or if you do a lot of turning between centers, it is advantageous to use a live center in the tailstock. Live centers are made with ball bearings so that they can rotate with the part (Fig. 6-4). This eliminates the need for constant oiling and makes turning between centers more pleasant.

Turning between centers has the advantage that no matter how many times you have to take your part out of the lathe, when you put it back in, it will always be perfectly centered. Also, because it is supported at both ends, there is less chance that your part will deflect away from the cutting tool. The disadvantage is that it is a little more work to set up because you have to drill the center holes at each end before you can start turning (Fig. 6-5).

Three-Jaw Chuck

The easiest method of holding work in a lathe is the *three-jaw chuck*. These chucks are made so that all three jaws move at the same time, and they automatically center round or hexagonal bar stock regardless of its size (Fig. 6-6). To use a three-jaw chuck, simply tighten the jaws on the part, then rotate the part either by hand or by running the lathe at its slowest speed to make sure the part is centered. When you have determined that the part is turning true, you can speed the lathe up and start cutting.

The problem with three-jaw chucks is that there are inaccuracies in even the best of them. They don't center a part perfectly. As long as you leave the part fixed in the chuck until it is completed, these inaccuracies probably won't affect you. If for some reason you have to temporarily remove the part, it may be as much as 0.004 inch or 0.005 inch off-center when you rechuck it.

You can minimize this error by marking the part next to the number 1 jaw on the chuck before removing it. When you rechuck the part, use the mark to align the part as close to its original position as possible. This method sometimes works, but if you know in advance that you will be chucking the part more than once, it may be best to use one of the other methods to hold your part.

Four-Jaw Chucks

The jaws-on four-jaw chucks move independently (Fig. 6-7). This means that you can center most objects quite accurately—even irregular shapes. You can also use four-jaw chucks to machine parts that have two or more cylindrical surfaces on offset axis. This includes parts like crankshafts and eccentric cams.

To mount your part in a four-jaw chuck, visually center the part in the chuck and bring the jaws in one at a time to grip it (Fig. 6-8). The part should be held firmly but not tightly. If the part is turned slowly, it can be marked with a pencil or a piece of chalk to indicate if it is off-center. Simply hold the chalk so that it just touches the side of your part as it rotates. To center the part, loosen the jaw opposite the mark and tighten the jaw on the same side as the mark. This process may need to be repeated several times. With a little patience you should be able to center a part to within 0.001 inch or 0.002 inch. Once the part is centered, tighten the jaws about one-eighth of a turn at a time until the part is gripped firmly enough so it won't slip or come loose while you are working it.

Should it be necessary to center the part even more accurately, you can mount a dial indicator on the tool post. Rotate the part by hand and watch the indicator to determine which way the part has to be moved to center it. The accuracy of centering a part this way is limited only by your patience and the accuracy of the dial indicator.

Collets

Collets provide a very accurate means of centering and holding small parts in the lathe. They are

Fig. 6-4. A plain center (right), when mounted in the tailstock, does not rotate with the workpiece and requires frequent oiling. The point of a live center (left) is bearing mounted to rotate with the workpiece. These centers do not require oiling.

used wherever precision centering is required. The collet is a hollow arbor. One end of the collet is tapered to match the taper in the lathe spindle. The same end has a precision bore in it and is slotted so that as the collet is drawn into the taper in the spindle, it can squeeze down and grip a part. The collet is drawn into the spindle by means of a drawbar that threads onto it (Fig. 6-9).

The advantage of using collets is that they do provide a fast, accurate means of centering and holding small parts. The disadvantage is that any one collet will only accommodate a very small range of diameters (Fig. 6-10). To have the capability of holding a large range of diameters, you need a lot of collets, and this gets expensive. For my shop, I have a set of five fractional size collets starting with 1/16-inch diameter and increasing in 1/16-inch increments up to 5/16-inch diameter. This permits

me to use collets for fractional size bar stock. The rest of my work is done either in a chuck or between centers.

BASIC LATHE OPERATIONS

The basic operations that can be performed on a lathe are facing, turning, lathe filing, parting, drilling, reaming, and boring (Fig. 6-11). You will find that each of these operations is quite easy to do, and you will be surprised at how quickly you can master them.

Facing

The term *facing* describes the process of machining a flat surface on a part 90° to the axis of the lathe. Parts to be faced can be mounted in the lathe by any conventional method. Chuck or collet mounting is preferred, but when required, the part

Fig. 6-5. Drilling a center hole in the lathe.

Fig. 6-6. All three jaws on a three-jaw chuck move at the same time and are considered self-centering. The centering accuracy is usually within .005 inches.

Fig. 6-7. The jaws on a four-jaw chuck move independently. These chucks can be used to center a part quite accurately. They are also used to hold irregularly shaped parts.

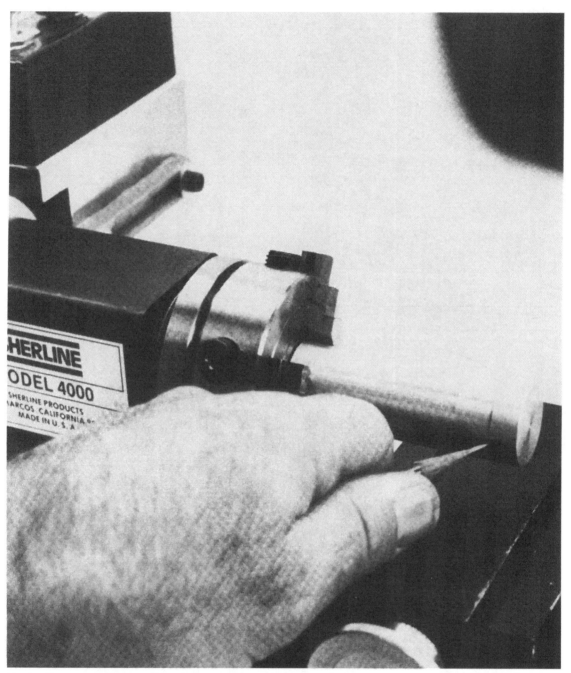

Fig. 6-8. To center a part in a four-jaw chuck, the part is first visually centered as close as possible. Then, with the lathe running slowly, the part is marked with a pencil or a piece of chalk to indicate if it is off-center. Final centering is accomplished by loosening the jaw opposite the mark and tightening the jaw adjacent to the mark. This process may have to be repeated several times.

Fig. 6-9. Collets are the most accurate chucks. They are used with a draw bar, which pulls the collet into a taper in the lathe's spindle. As the collet is drawn into the taper, it squeezes down to grip the part.

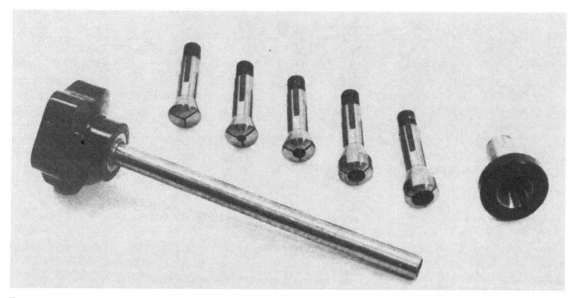

Fig. 6-10. Each collet has a very limited range of diameters that it can accommodate. A complete set of collets to accommodate a wide range of diameters would be quite expensive. This set of Sherline collets consists of a draw bar, a collet adapter, and five collets.

can be mounted between centers. The only problem with facing a part held between centers is that you need a specially shaped lathe tool (Fig. 5-11), and you won't be able to eliminate the center hole.

For facing, the height of the cutting tool should be set as close to the turning center as possible. This adjustment can be made simply by placing a center in the tailstock and adjusting the tip of the cutter to the tip of the center. The tool should also be set into the tool holder as deeply as practical to minimize the tendency to chatter. When the tool has been secured in place, the tool post can be rotated into the proper position for facing.

Start your cut at the center of rotation and work to the outside of the part on the side closest to you. The depth of the cut and the feed rate will vary depending on the material, the size of the part, and the size of the lathe. To get a nice finish, I like to make the final cut a light one, no more than 0.004 inch to 0.005 inch deep. The feed should be turned as slowly and as smoothly as possible.

Turning

Making a cut along the outside diameter of a part is called *turning*. For turning, the part can be mounted in the lathe by any conventional method, chuck, collet, or between centers. Most turning is done by cutting from right to left (Fig. 6-12). It may be preferable to go from left to right for some parts. The lathe tool should be mounted with its nose set on center or slightly below center. The tool should be set as deeply into the tool holder as possible.

As with facing, the depth of cut and feed rate will depend on the type of material you are cutting, the size of the part, and the size of the lathe. I don't like to overstress my machinery, so if I start a cut and the lathe seems to be strained, I back off and try a smaller cut. Remember that a 1/32-inch deep cut will reduce the diameter of your part by 1/16 inch. It doesn't take long to remove a lot of stock.

On most lathes, the crossfeed is calibrated to indicate the depth of cut in thousandths of an inch. If you determine that you want to reduce the diameter of a part by some dimension, say 0.020 inch, simply turn the crossfeed in by half that amount (0.010 inch) and start your cut.

You will find a certain amount of play in the feedscrew of any machine. For this reason the depth

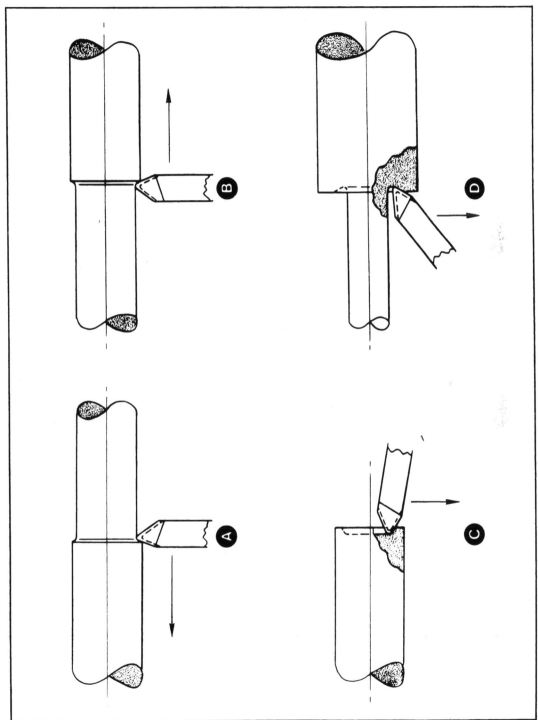

Fig. 6-11. The basic lathe operations are: (A) right-hand turning, (B) left-hand turning, (C) right-hand facing, (D) left-hand facing. The arrows indicate the direction of feed.

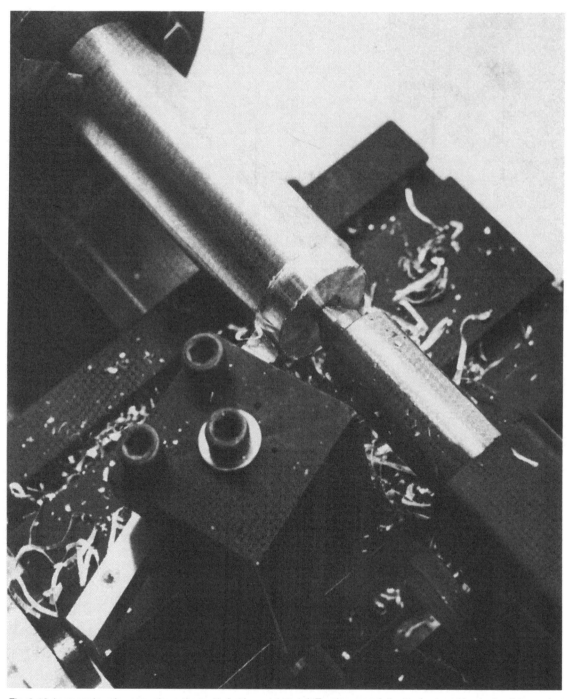

Fig. 6-12. In normal turning operations, the tool is fed from right to left. Turning speed, depth of cut, and feed rate all depend on the type and size of material being cut. Note the use of a center in conjunction with a three-jaw chuck.

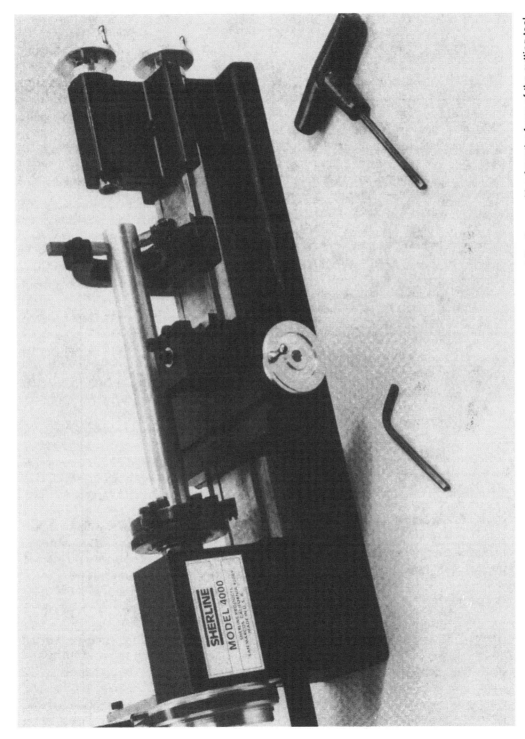

Fig. 6-13. A steady-rest is used to support long, slender parts for center drilling and prevent part deflection resulting from the force of the cutting tool.

of cut should always be adjusted by moving the tool toward the part. If for some reason you decide to take a lighter cut, back the tool out further than you need to and then bring it back in for the correct depth of cut. This will take up the play in the feedscrew. This rule applies to facing operations as well as turning operations.

The turning speed will vary depending on the material and the diameter of the part. The drilling charts in Chapter 4 can be used as a turning speed guide. The speed is not critical, and you may want to experiment a little. Just remember that soft materials and small diameters can be turned faster than hard materials and large diameters. Slow speeds create no special problems; it just takes longer to finish the part. Excessively fast speeds create too much heat, and the cutting tool will lose its cutting edge. Cutting oil may be used when cutting steel to help prevent overheating. Cutting oil should always be used when cutting aluminum to prevent a buildup of material on the cutting tool.

One problem frequently encountered in lathe work is deflection of the workpiece. This tendency is especially noticeable in long slender parts. The result of this deflection is that you will end up with an uneven diameter on your part. Light cuts and slower feed rates will help to reduce the problem, but they won't solve it. The easiest way to solve the problem is to add support to the part. If the part is supported at one end only, support the free end with a center or a *steady rest* (Fig. 6-13). If the part is already supported at both ends and it still deflects, use a steady rest to support the center. If these remedies fail, you may find that the easiest way to finish the part is by lathe filling.

Lathe Filing

Lathe filing (Fig. 6-14) is used not only to correct the tapers that result from cutting tool pressure, but is a technique that can be used to form radii and otherwise shape parts when it is impractical to grind a special tool to machine these shapes. As a rule, fine single-cut files are used. *Caution: lathe filing can be dangerous. Always wear your safety*

glasses and never attempt to lathe file with a file that does not have a good handle.

The lathe should be turning approximately 50 percent faster than the normal turning speed. Hold the file with both hands. Keep it at a slight angle and use long steady strokes with light downward pressure. Check the file frequently for clogging. Check the part to make sure you don't remove too much material.

Cutting Chamfers and Turning Tapers

Facing and cylindrical turning are controlled by the main feedscrew and the cross-slide feedscrew. In addition to these two feeds, most of the larger lathes have a third feed called the *compound feed*. This feed can be rotated in various ways to cut almost any kind of chamfer or taper on the part. As a rule, the compound slide is mounted on a protractor so that you can set it with a fair degree of accuracy for almost any angle.

On some of the smaller lathes like the Sherline, these operations are done a little differently. Small chamfers are cut by rotating the tool post so that the side cutting edge of the tool will cut the desired angle (Fig. 6-15). For cutting tapers the headstock can be rotated so that it holds your workpiece at an angle to the bed of the lathe (Fig. 6-16). The Sherline headstock is set on a protractor to facilitate setting it up for almost any desired angle.

For cutting accurate tapers, the final adjustment needs to be done by trial and error regardless of whether you have a lathe with a compound slide or a lathe with a rotating headstock. As an example, suppose that you wanted to cut a #1 Morse taper. A #1 Morse taper goes from a diameter of 0.517 inch to a diameter of 0.367 inch in 3.000 inches. The difference between the two diameters is 0.150 inch $(0.517 - 0.367 = 0.150)$.

One-half the difference is 0.075 inches. Knowing this, chuck up a piece of drill rod or ground bar stock at least 3.000 inches long. Using the cross-slide feed, bring your tool in until it just touches the outside diameter of the bar stock at the free end. The height of the tool must be set on the

Fig. 6-14. Lathe filing is used to correct tapers resulting from deflection of the part away from the lathe tool. Lathe filing is also used to form radii or other shapes when it is impractical to grind a special tool to machine these shapes.

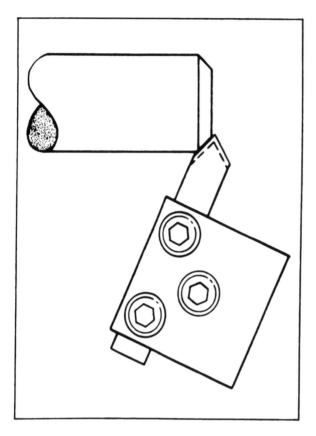

Fig. 6-15. Small chamfers are machined by rotating the tool so that the side cutting edge of the tool will cut the desired angle.

turning center of the lathe. Next, bring the tool over exactly 3.000 inches toward the headstock. Then using the cross-slide, bring the tool in again until it just touches the part. The distance you move the tool can be read directly on the cross-slide control knob. When that distance is exactly 0.075 inch your lathe is set up for turning a #1 Morse taper. This same procedure can be used to set your lathe up for cutting any taper.

Parting

The operation of separating a part from excess stock is called *parting*. The tool used for this operation is a long slender blade, as described in Chapter 5. When setting your lathe up for a parting operation, set the height of the cutting tool as close to the turning center as possible (Fig. 6-17). The tool holder should be rotated so that the blade of the parting tool is parallel to the cross-slide's direction of travel.

Parting should only be done with the part mounted in a chuck or collet. It should never be done with the part mounted between centers. If the part were mounted between centers, it would start to bend under the pressure of the cutting tool. This would cause the freshly cut surfaces of the part to pinch the tool. The result could be damage to the part, damage to the lathe, or a broken parting tool.

For parting operations the turning speed should be reduced to one-third or less the normal turning speed (Fig. 6-18). The tool should be advanced slowly and smoothly. Tool chatter is generally an indication that the turning speed is still too fast. When you get close to completing the cut, the part, if not too big, can be held loosely in your hand so that it will not fly off when it separates. Caution:

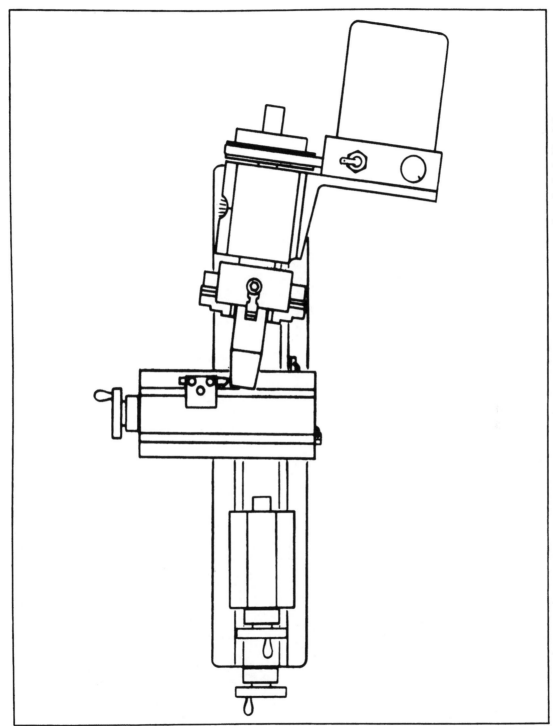

Fig. 6-16. On the Sherline lathe, tapers are cut by rotating the lathe's headstock to hold the work at an angle to the bed of the lathe.

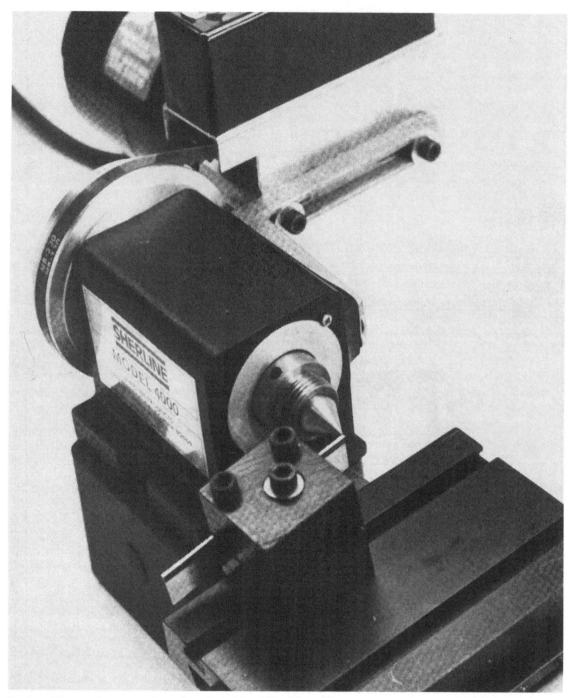

Fig. 6-17. When setting up a parting tool, set the height of the cutting edge as close to the center of rotation as possible. This adjustment is best made using the tip of a center as shown here.

114

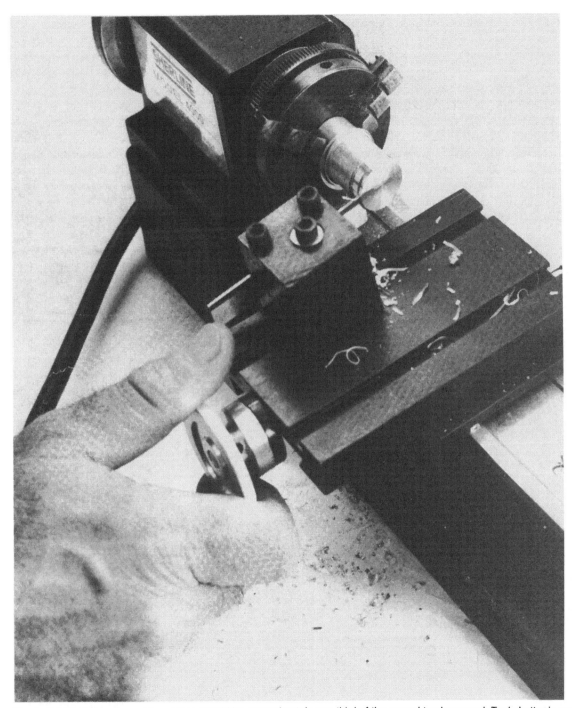

Fig. 6-18. The turning speed for parting should be approximately one-third of the normal turning speed. Tool chatter is an indication that the turning speed is too fast.

use extra care when working with parts that have sharp edges.

Drilling and Reaming

The lathe makes an excellent drill press when you want a part drilled exactly on its center. The part to be drilled normally is held in the headstock with either a chuck or a collet. The drill is mounted in a drill chuck in the tailstock. The tailstock is brought up so that the drill is almost touching the part, then it is locked in place. The drill can then be advanced into the part by means of the tailstock ram (Fig. 6-19). Use plenty of cutting oil for most materials. During the drilling process the drill should be removed frequently to clear the chips.

When drilling small diameter holes, use a center drill first. Small diameter drills have a tendency to flex and you may end up with an off-center hole or a broken drill. The center drill has a heavy shank and it won't flex. If you start the hole with one, it should be right on center.

If the part you are working is very long, support the free end with a steady-rest to make sure it is turning true. To set up a steady-rest, first mount the part in a chuck or collet. Then set the steady-rest on the bed of the lathe, slide it over the free end of the part until it is as close to the headstock as possible, and clamp it in place. Now you can adjust the three blades on the steady-rest so that they just touch the part. When the blades have been adjusted, you can loosen the steady-rest clamp and slide the steady-rest back to support the free end of the part.

Fig. 6-19. When drilling in a lathe, the drill is fed into the part by means of the tailstock ram. The drill should be removed frequently to clear the chips.

Fig. 6-20. When reaming a part in a lathe, the reamer can be fed into the part by pushing on the tailstock with your hand.

Be sure to clamp the steady-rest in place when you get it located where you want it. Before starting the lathe, you will need to apply a little oil where the blades touch your part.

Reaming operations are done almost the same as drilling, except that you don't need to lock the tailstock in place and advance the reamer with the tailstock ram. Instead, you can feed the reamer into the part by just pushing on the tailstock with your hand (Fig. 6-20). Remember, for reaming, turning speeds and feed rates should be slow. You should use plenty of cutting oil to obtain a good finish.

Boring

Lathe boring (Fig. 6-21) is a method for making large diameter holes, close tolerance holes, or specially shaped holes. It differs from drilling in that it uses a single-point cutting tool mounted in the tool holder. The part to be bored is usually mounted in a chuck or collet and supported by a steady-rest when required.

A drill is usually used to make a pilot hole. The pilot hole should be as large as practical. The larger the pilot hole, the heavier and stronger the boring tool can be. The boring tool can be ground from a standard lathe tool bit or, if you are making a deep hole, a boring bar (Fig. 6-22). When you grind your tool, make sure you grind enough clearance for the smallest diameter in which you will be working, but don't remove any more material than necessary. If

Fig. 6-21. Lathe boring is a method used to produce close tolerance or large diameter holes. It is also the method used for cutting internal grooves such as those required for retaining rings.

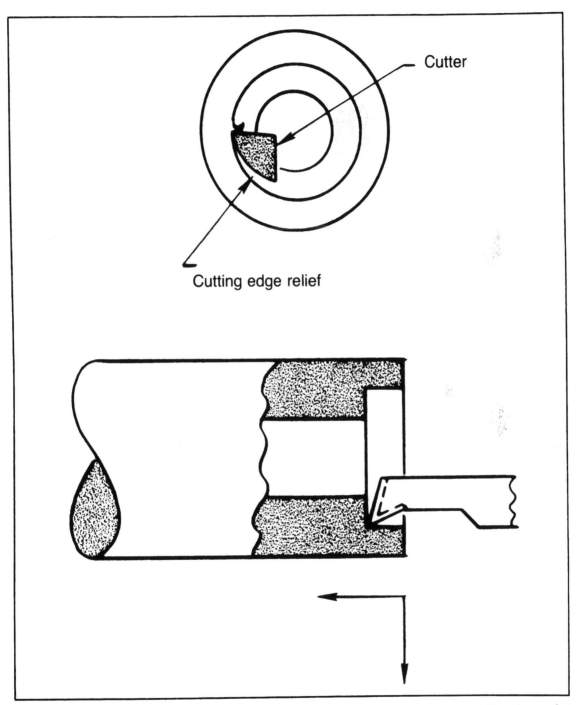

Fig. 6-22. You can grind your own boring tool from a standard lathe tool if the desired hole is not too deep. Just make sure that you grind enough clearance (cutting edge relief) for the smallest diameter hole that you want to machine.

the tool is ground too slender, it may chatter or break.

The rest of the boring procedure is very similar to turning. The cut is made on the side of the part closest to you. The direction of cut is normally right to left. Light cuts are recommended, and remember that a 0.010-inch deep cut will increase the diameter by 0.020 inch.

Threading

The easiest way to cut threads is with a tap or die because there is no worry about thread form or tolerances. As long as the thread is a standard one, the tap or die will cut it to the correct size and shape. There are times when it is necessary to deviate from thread standards, however, and then it is necessary to find a different way to cut them. Most metal working lathes provide that capability. Thread cutting is accomplished by gearing the headstock spindle to the feedscrew so that the tool will be advanced in direct relation to the rotation of the spindle. By varying the gear ratios, any desired pitch of thread may be cut. As an example, if the gear ratio is such that it moves the tool 1/10 of an inch (0.100) for every revolution of the spindle, the resulting thread will have a pitch of 10 (10 threads per inch). By changing the gearing to cause the tool to move 1/16 of an inch (0.062) for every revolution of the spindle, the resulting thread will have a pitch of 16 (16 threads per inch).

Not all lathes have threading capability, but those that do usually use one of two methods to change the gear ratio (Figs. 6-24 and 6-25). In the first method, the lathe is supplied with a set of gears that can be assembled by the operator in various combinations to obtain the desired thread pitches. Usually the various gears are all clearly identified, and the set comes with complete instructions on how to assemble them.

In the second method, the lathe comes with a built-in quick change gear box. The gear box can be compared to the transmission in a car. By manipulating one or more shift levers, you can obtain a number of different gear ratios. Obviously, lathes that have quick change gears have the advantage of convenience, but they are also more expensive. In the home shop where production speed is not important, the manually changed gear sets do a fine job.

Because the actual method of cutting threads varies with different lathe models and manufacturers, I will leave detailed thread cutting instructions to the manufacturer's instruction book and concentrate on some basics.

Thread standards were discussed in some detail in Chapter 3, but *thread form* was not covered. By international agreement, a standard thread form was established in 1948 for use on all American and English threads. It closely resembles thread standards for metric machine screws (Fig. 6-27).

For the type of work most of you will be doing in your home workshop, you only need to remember to use a 60° tool and to make sure that the tool has a generous side relief. The relief is necessary, especially when cutting coarse threads, to ensure that there is clearance between the face of the tool and the thread. The nose of the tool can be radiused slightly to form a small radius at the bottom of the cut. The purpose of the radius is to minimize high stress areas in the screw that may tend to weaken it. The size of the radius is not critical as long as it isn't so big that it prevents the thread from going into its mating part.

When setting the tool in the lathe, the tool height should be exactly on center. The tool must be set square with the work so that the thread angle will be correct. The easiest way to do this is with a threading gauge. Whenever possible, check the finished thread by screwing on the mating part before you remove your work from the lathe.

SAFETY

Your lathe is a very valuable tool. Lathes, regardless of their size, can be dangerous if misused. Always observe the following safety rules.

● Never wear loose clothing when operating a lathe.

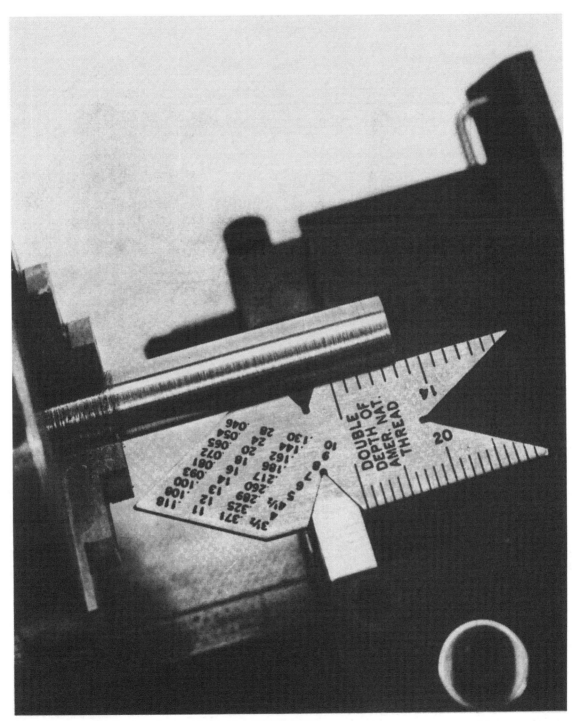

Fig. 6-23. A center gauge can be used to square a threading tool.

Fig. 6-24. The threading attachment for a Sherline lathe permits cutting 31 different unified thread pitches and 28 different metric thread pitches.

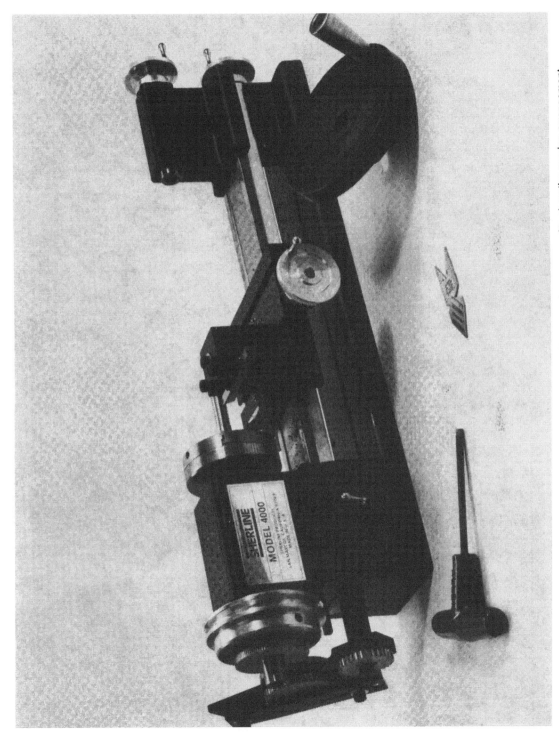

Fig. 6-25. A Sherline lathe set up for cutting threads. The large handwheel in the foreground was removed to show the gearing arrangement.

Fig. 6-26. Cutting a 16-pitch thread with the Sherline lathe.

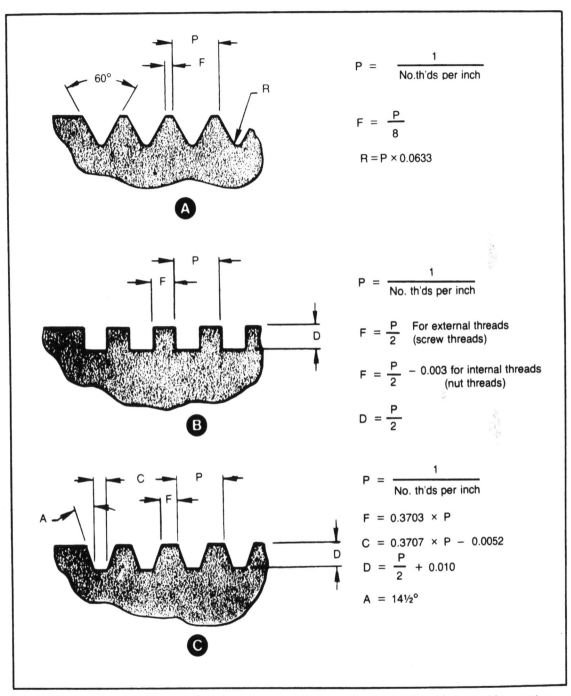

$$P = \frac{1}{\text{No.th'ds per inch}}$$

$$F = \frac{P}{8}$$

$$R = P \times 0.0633$$

$$P = \frac{1}{\text{No. th'ds per inch}}$$

$$F = \frac{P}{2} \quad \text{For external threads (screw threads)}$$

$$F = \frac{P}{2} - 0.003 \text{ for internal threads (nut threads)}$$

$$D = \frac{P}{2}$$

$$P = \frac{1}{\text{No. th'ds per inch}}$$

$$F = 0.3703 \times P$$

$$C = 0.3707 \times P - 0.0052$$

$$D = \frac{P}{2} + 0.010$$

$$A = 14\frac{1}{2}°$$

Fig. 6-27. The three most common standard thread forms are: (A) International screw thread form, used for most fastener threads, both metric and English; (B) Square screw thread form, used for vise screws, jack screws, and machine feed screw threads; (C) Acme screw thread form used for machine feedscrew threads.

● Remove all jewelry before operating a lathe. This includes rings, watches, identification bracelets, etc.

● Always wear safety glasses.

● Keep your hands clear of moving parts.

● Get in the habit of removing chuck keys and chuck tools. Don't take your hand off them unless you remove them.

● Never turn a lathe on unless the chuck jaws have been tightened.

● Never remove chips with your bare hands.

● Never try to use a micrometer or caliper on a part while the lathe is running.

● Use common sense.

Chapter 7

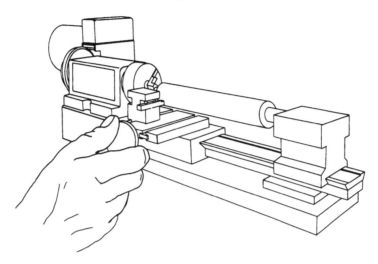

Milling Machines

IF THE LATHE WAS THE MOST IMPORTANT STEP forward in the development of machine tools, then the milling machine must be the second most important. These machines are used for machining flat surfaces, profiling, slitting, slotting, hollowing out, precision drilling, and boring operations. There are two basic types of milling machines: the horizontal milling machine and the vertical milling machine (Fig. 7-1). The terms horizontal and vertical refer to the normal orientation of the cutter axis.

On horizontal milling machines, the cutter is mounted on a horizontal arbor. Usually the arbor is supported at both ends, resulting in greater rigidity. These machines are capable of removing a lot of material in a short time, but horizontal milling machines are not as versatile as the vertical milling machines and are usually found only in production shops.

Vertical milling machines (Fig. 7-2) normally rotate their cutters around a vertical axis. On most machines the cutter axis can be rotated away from the vertical, enabling the machine to make angular cuts. As a rule, the cutter and arbor are one piece.

They are collet-mounted into the machine. The cutter is supported from one end only. Because the vertical machines are the most versatile, they are very well suited for work in the small machine shop. Therefore, most of this discussion will be regarding them, but it should be remembered that the same principles do apply to both types.

On a milling machine, the work is mounted on a bed or table directly under the cutter. Then, by means of feedscrews, the relationship between the work and the cutter can be accurately controlled on three axes. Because of its convenient size and versatility, the Sherline vertical milling machine, will be featured in this chapter. The only differences between this small machine and the large ones are size and the fact that most of the larger machines control the Z axis by raising and lowering the table. The Sherline accomplishes this function by raising and lowering the milling head.

SELECTING A MILLING MACHINE

In looking for a milling machine to purchase, your first consideration should be size. You want to

127

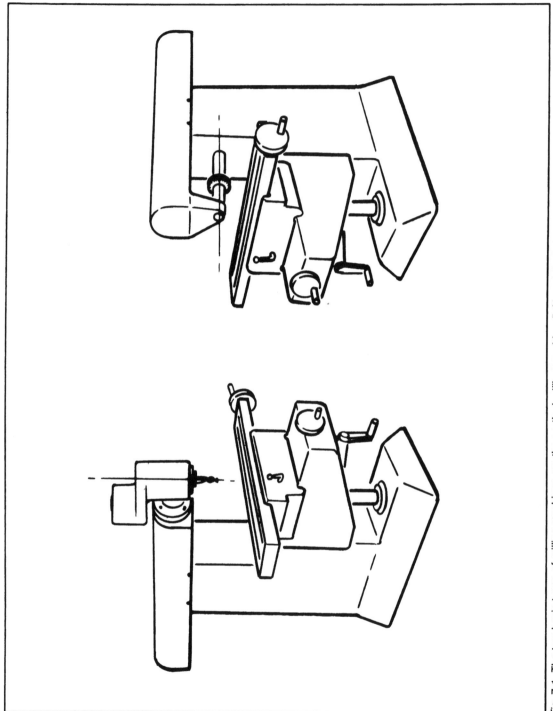

Fig. 7-1. The two basic types of milling machines are the vertical milling machine (left) and the horizontal milling machine (right).

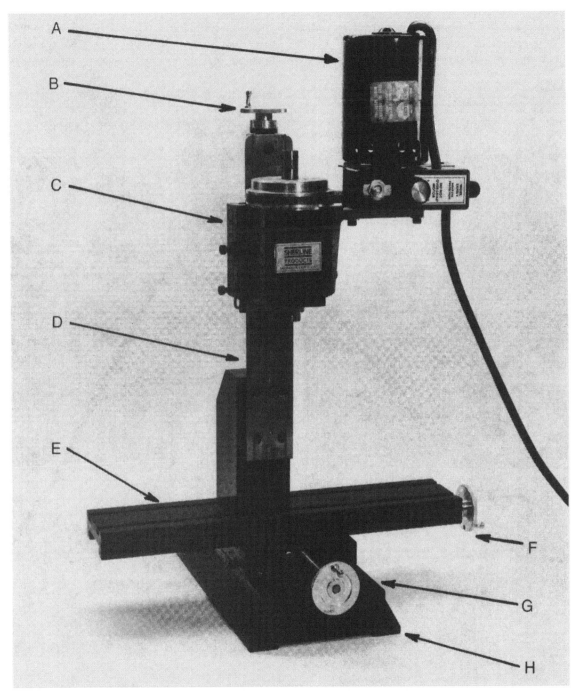

Fig. 7-2. The vertical milling machine is the most versatile of all machine tools. The major parts of a vertical milling machine are: (A) motor, (B) Z axis feedscrew, (C) milling head, (D) vertical column, (E) table, (F) X axis feedscrew, (G) Y axis feedscrew, and (H) base.

make sure that the machine you purchase will be large enough to accommodate the type of projects on which you want to work. The dimensions that you will want to know are table size, travel on both the X and Y axes, maximum spindle height above the table, and the distance between the spindle axis

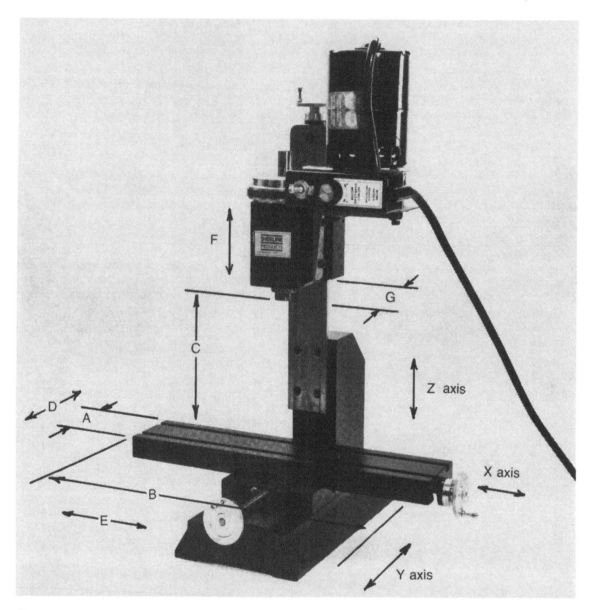

Fig. 7-3. The three axes of movement on any milling machine are the X, Y, and Z axes as identified here. The important dimensions to consider when selecting a milling machine are: (A) width of table, (B) length of table, (C) maximum spindle height above the table, (D) travel on Y axis, (E) travel on X axis, (F) travel on Z axis, (G) distance between spindle and vertical column.

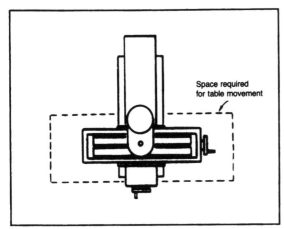

Fig. 7-4. When determining the amount of shop space required for a milling machine, it is important to allow for full movement of the machine's table.

to increase with the size of the machine. In selecting a milling machine, resist the temptation to purchase a machine larger than you actually need, not only for the obvious reasons of cost and shop space, but also because it is easier to do fine work on a small machine.

When you have determined the size machine you want, compare available features. Rigidity in a milling machine is even more important than it is in a lathe. Milling machines should be ruggedly built. The slides should have adjustable gibs so that you can compensate for wear, and they should be kept tight at all times. There should be provisions for tightening spindle bearings. Finally, the feed-screws should be shielded from chips produced during the operation of the machine.

CUTTERS

One thing that makes the vertical milling machine so versatile is the number of different cutting tools that it can use. The most common type of cutter is the *end mill* (Fig. 7-5). End mill cutters are designed specifically for vertical milling machines. They are one piece, complete with their own arbor. These cutters are used extensively for facing, plunge cutting, hogging out cavities, and contour milling.

Like most cutting tools, end mills are available in both high speed steel and carbide. They are made with two, three, four, or more cutting edges and in many sizes.

and the support column. Refer to Fig. 7-3 to better visualize these dimensions. These will determine the size of the parts you can make.

In determining the size of the machine you want to purchase, you will also want to consider how much shop space it will require. Remember that in addition to the overall size of the machine, you need space to permit the table to travel back and forth (Fig. 7-4). This means that the width of the space you will require will be equal to the width of the machine plus the maximum travel on the X axis.

The last consideration is cost. Naturally, the larger the machine the greater the initial cost will be. The cost of accessories and operation also tends

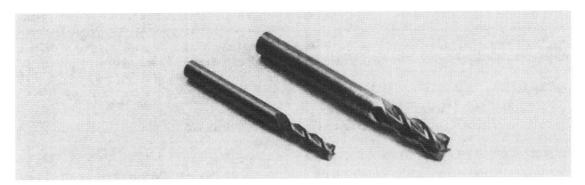

Fig. 7-5. The cutters most commonly used in a vertical milling machine are called end mills. The cutters shown here are four-flute, high speed steel end mills.

When selecting an end mill, your first consideration should be material—either high-speed steel or carbide. I prefer the high speed steel cutters. Carbide cutters are more expensive and are quite brittle. Carbide cutters are preferred for production machining of hard materials on automatic machines, but for the kind of work most of you will be doing in your small workshops, high speed tools do a fine job. There is less chance of breakage.

The next consideration in selecting an end mill is size. You generally should select the largest diameter cutter that you can use for a job. Large diameter cutters are stronger, so there is less chance of breakage. There is less flexing because they are stronger, so it is easier to hold tolerances. Still another benefit is that the large cutters have larger flutes for more chip clearance. These advantages permit heavier cuts, faster feed rates, and more precision. The only time you need a small diameter cutter is when you are cutting a narrow slot or cutting out a pocket that requires small corner radii.

When cutting pockets, the diameter of the cutter you are using will determine the size of the corner radius in the part. A *corner radius* is the radius left between any two adjacent sides of the pocket (Fig. 7-6). A ½-inch diameter cutter will leave a ¼-inch radius; an ⅛-inch diameter cutter will leave a 1/16-inch radius. If a very small corner radius is desired, it is generally preferable to use a large diameter cutter to rough out the pocket, and then a small diameter cutter to finish cut the pocket and cleanup (cut to size) the corner radius.

Whenever it is desirable to leave a radius between the side wall and the bottom of a pocket, you will want to use a *ball end mill*. The difference between a ball end mill and a conventional end mill is illustrated in Fig. 7-7. The disadvantage of using a ball end mill is that it is hard to machine a good flat surface at the bottom of a pocket with one. For this reason a standard cutter is usually used to rough cut the sides of the pocket and finish cut the bottom. Then a ball end mill is used to finish cut the sides and the bottom radius.

The last consideration when selecting an end mill is the number of flutes, or cutting edges, the cutter should have. The number of cutting edges is always the same as the number of flutes (Fig. 7-8). Because it is easier to say flutes, it has become customary to talk about the number of flutes rather than the number of cutting edges. End mills are available with two, three, four, or multiple flutes (five or more). The number of flutes best suited for you depends on the type of work you are doing.

Two-flute cutters are considered good roughing and general purpose cutters. They are used primarily for soft materials such as brass and aluminum. The large flute size gives plenty of room for large chips so heavy cuts and fast feed rates can be taken. Two-flute cutters can be used for both plunge cutting and side cutting. In plunge cutting, the cutter is fed into the work end first just as a drill is fed into a part. Plunge cut feed rates are slower for an end mill than they are for a comparable size drill. Also, there is more tendency for an end mill to wander off-center. Thus, end mills make poor substitutes for drills.

The disadvantage of two-flute cutters is that for finish work they have to be fed slower than cutters with more flutes. Figure 7-9 illustrates the reason for this difference. Another disadvantage of two-flute cutters is vibration. As a rule, the more flutes a cutter has, the less chance there is for unwanted vibration that makes it difficult to hold good tolerances and obtain a good finish.

Four-fluted end mills are good all-around cutters. They can be used on all types of materials from very soft to very hard. Because of the reduced flute size, they are not as well suited for roughing operations on soft materials. Because the feed rates are slower, however there is plenty of chip clearance when roughing harder materials. Another advantage of the four-fluted cutter is that the two added cutting edges give the cutter more rigidity.

Some four-fluted cutters cannot be used to make plunge cuts (Fig. 7-10). It depends on the way they have been ground. Before attempting a plunge cut with a four-flute cutter, examine the end cutting

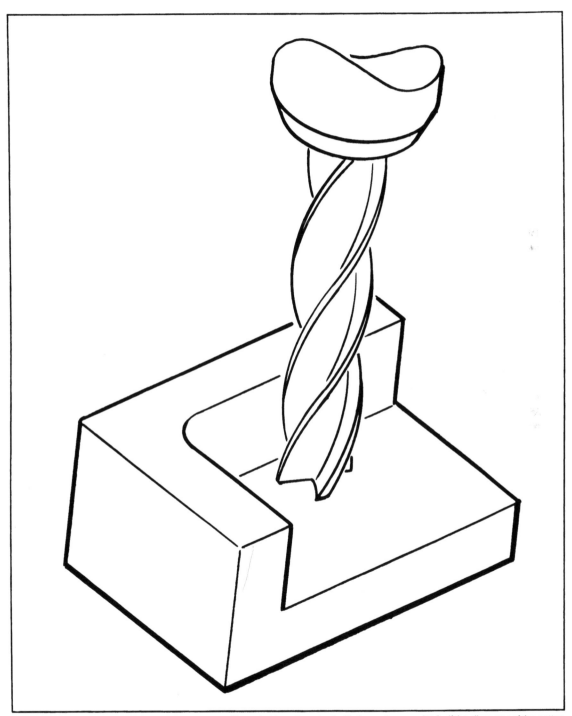

Fig. 7-6. The corner radius left between any two adjacent sides of a cutout will always be equal to half the diameter of the cutter.

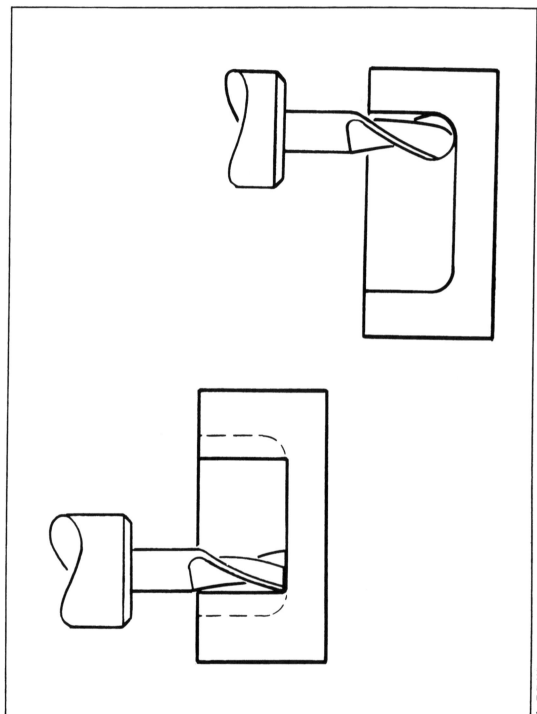

Fig. 7-7. When it is desirable to leave a radius between the sides and the bottom of the cut, you will need to use a ball end mill. In these cases, the cutout is first "roughed out" with a standard end mill and then finished using the ball end mill.

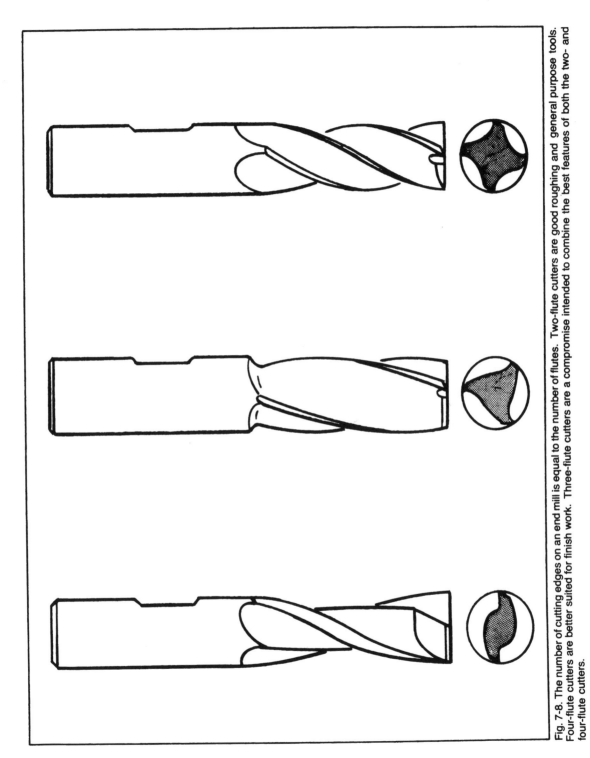

Fig. 7-8. The number of cutting edges on an end mill is equal to the number of flutes. Two-flute cutters are good roughing and general purpose tools. Four-flute cutters are better suited for finish work. Three-flute cutters are a compromise intended to combine the best features of both the two- and four-flute cutters.

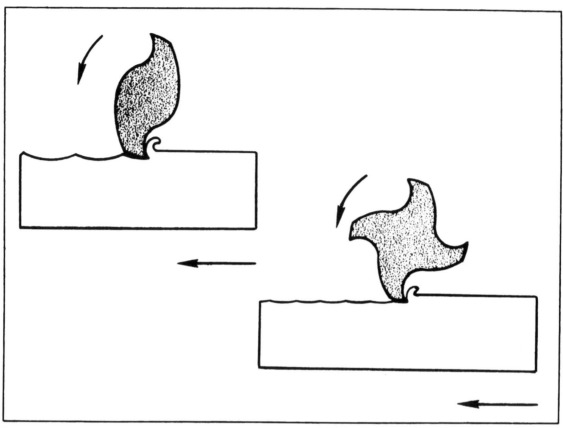

Fig. 7-9. Given the same cutter speeds and feed rates, the surface finish left by a four-flute cutter will be much smoother than the finish left by a two-flute cutter.

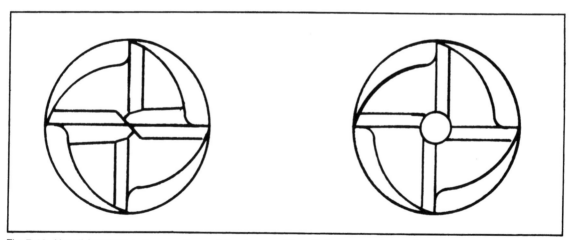

Fig. 7-10. Not all four-flute cutters can be used for plunge cutting. Before attempting a plunge cut, examine the cutter to make sure that at least two of the cutting edges extend all the way to the center of the cutter.

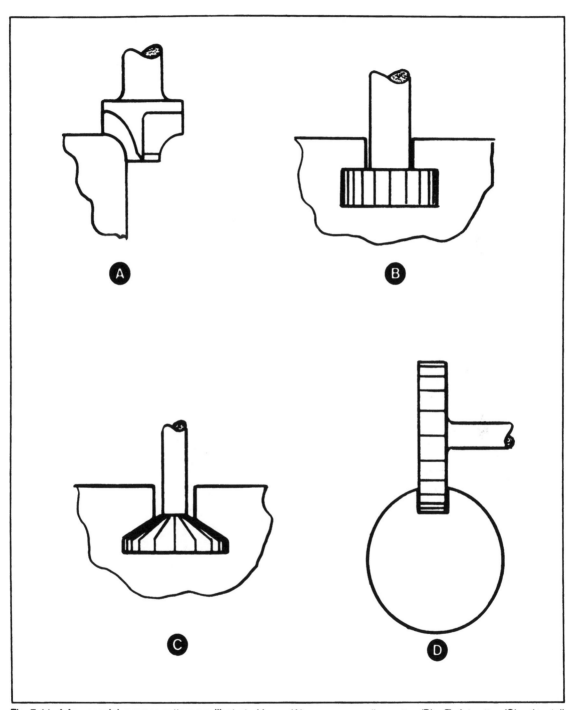

Fig. 7-11. A few special purpose cutters are illustrated here: (A) a corner rounding cutter, (B) a T slot cutter, (C) a dovetail cutter, and (D) a Woodruff keyset cutter.

Multiple-flute end mills may have five or more flutes. They are intended for close tolerance finish work, because with their large center core, vibration and deflection are reduced to a minimum. Multiple-flute cutters are seldom ground to permit plunge cutting.

Special End Mill Cutter. Special end mills are available for cutting special shapes. Figure 7-11 illustrates a few. The corner rounding cutters, T slot cutters, dovetail cutters, and Woodruff Keyset cutters are the most common ones. All of these cutters are available in different sizes.

Shell End Mill Cutters. Shell end mill cutters (Fig. 7-12) are generally used only on the larger production milling machines. It's doubtful if anyone working in a small shop will have a need for one of these cutters, but a brief discussion may prove of some value. The word *shell* indicates that

Fig. 7-12. Shell end mills are end mills with removable arbors. This style of end mill is available only in the larger sizes, usually 1¼-inch diameter and larger.

edges. Make sure that at least two of them go all the way to the center of the cutter.

Three-flute end mills are designed to combine the best features of two-flute and four-flute cutters. They can be used for both side cutting and plunge cutting. These cutters are excellent all-purpose tools.

Fig. 7-13. Most circular cutters have cutting edges on the sides as well as on the circumference. These cutters are used primarily on horizontal milling machines.

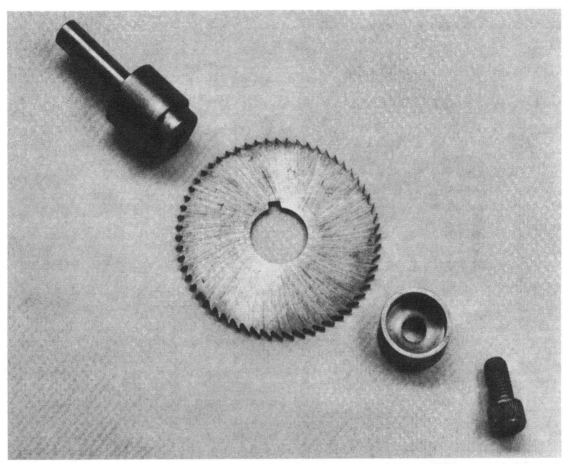

Fig. 7-14. Unlike circular cutters, slitting saws have no side cutting edges and should be used for sawing only. This one is shown with an arbor.

these cutters are made with a removable arbor. They are only made in large sizes, usually 1¼-inch diameter and bigger. The advantage of making these large cutters with a removable arbor is strictly economical. Tool steel is expensive, and if you can use one arbor to accommodate several different sizes of cutter, you can save some money.

Circular Cutters. Circular cutters (Fig. 7-13) are used primarily on horizontal milling machines, but they can also be used on vertical machines. They are made with a keyed center hole for arbor mounting. These cutters are made in various diameters and widths, with a minimum width of approximately 1/16 inch. Most of these cutters have cutting edges on both sides as well as on the circumference. They are strong enough to withstand side loads and can be fed into a cut sideways.

Circular cutters are frequently ganged together on an arbor and used for cutting thin walls. If you cut a wall with an end mill, you can cut only one side at a time. If the wall is a thin one, it may deflect away from the cutter. You will end up with varying wall thickness. With ganged circular cutters you cut both sides of the wall at the same time, so there is no deflection. It is easier to maintain a uniform wall thickness.

Like end mills, circular cutters are also available in a number of special shapes. Such as bevel

Fig. 7-15. A slitting saw mounted on its arbor.

cutting cutters, corner rounding cutters, and gear tooth cutters.

Slitting Saws. Although slitting saws (Fig. 7-14) look very similar to circular cutters, the two should not be confused. As the name implies, these cutters are intended for sawing and sawing only (Fig. 7-15). They do not have cutting edges on either side, and they are not designed to withstand side loads. These cutters are intended for cutting thin slots. They come in various widths starting at 0.006 inch and increasing in width up to 3/16 inch. The thinner ones are usually about 1 inch in diameter, and the diameter increases as the width increases (Fig. 7-16).

Fly Cutters. Fly cutters can be used in vertical milling machines. They make short work of cutting large flat surfaces. These tools are single-point cutters using an inserted cutter similar to a lathe tool. Before using one of these tools, the milling machine head should be checked to make sure it is square to the bed of the milling machine. If

it is not, the cutter will cut shallow scallops in the workpiece (Fig. 7-18). Once the machine is set up, the rest is easy. Just feed the cutter across the part taking a fine cut. The result is a good flat surface with a nice finish.

Boring Tools. It is not feasible, even for a large commercial machine shop, to own complete sets of reamers so that any size hole can be accurately machined. Also, there are times when it is necessary to machine a close tolerance hole to a configuration where a reamer won't work. With a boring tool, any size or configuration hole can be accurately machined. A boring tool is a single-point cutting tool that can be adjusted with micrometer accuracy. Any diameter hole can be bored within the limits of its adjustment. Like the fly cutter, this tool uses an insert cutter similar to a lathe tool. By means of a graduated lead screw, the cutter can be positioned for cutting various diameter holes (Fig. 7-19).

Boring tools are never used to start a hole. The hole is always predrilled undersize, then the boring tool is used to open it up to its required size. Usually the cutter is positioned to take one or more light cuts. When the hole is almost to size, it is measured and the boring tool is readjusted to make the finish cut. These tools are quite handy, and they soon pay for themselves.

COLLETS

Because end mills typically have more than one cutting edge, it is important that the cutter be centered in the spindle so that it will run as true as possible or the cutter will work unevenly. One cutting edge will take a deep cut while the opposite cutting edge will take a light cut. The result of this unevenness would be a shortened cutter life and a poor finish on your work. For these reasons, end mills should always be collet-mounted (Fig. 7-20). Collets that are used with a milling machine are usually a little different from those used for holding a part in a lathe. The taper is more gradual, resulting in greater holding power. These collets are a

Fig. 7-16. A slitting saw in action.

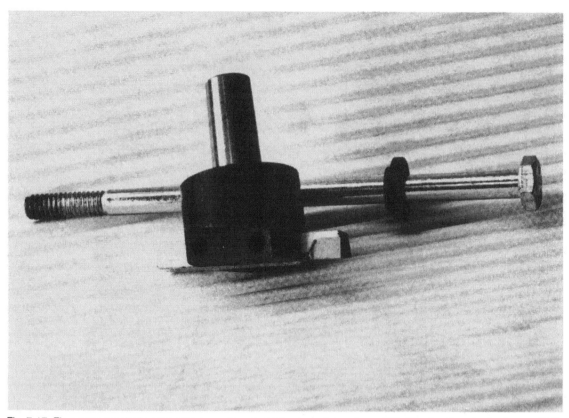

Fig. 7-17. Fly cutters are single point cutters used for machining large flat surfaces. This Sherline fly cutter is shown with its draw bar.

necessary accessory for milling machines and should be purchased with the machine if possible.

SETUP TOOLS AND ACCESSORIES

Good setup tools and accessories are just as important to good milling operations as good cutters. These tools and accessories include T nuts, T bolts, clamps, parallel bars, angle plates, sine bars, V blocks, and machine vises. When you first set up your shop, you won't need all of these tools. You can get started with just a few basic items and add to them as the need arises. You might even want to use your machinery and make some of these accessories yourself. Plans for some of them are shown in Chapter 11.

T Nuts and T Bolts. The tables of most milling machines are made with T slots. These run the length of the table and are used for holding the nuts and bolts that are used to clamp your work to the table. T nuts and T bolts are made to fit into the T slots. The slots prevent them from turning as they are tightened or loosened. These nuts and bolts are also used to hold most accessories such as clamps, angle plates, and vises to the milling table. They are essential to the operation of a milling machine. Every shop should have an ample supply (Fig. 7-21).

Clamps. The most common and versatile type of clamp is the *strap clamp* (Fig. 7-22). A strap clamp is an iron or steel bar and two bolts. One bolt is threaded into the back end of the clamp and is adjusted to hold the clamp in a horizontal position when the other end of the clamp is resting on the part. The second bolt passes through a hole in the

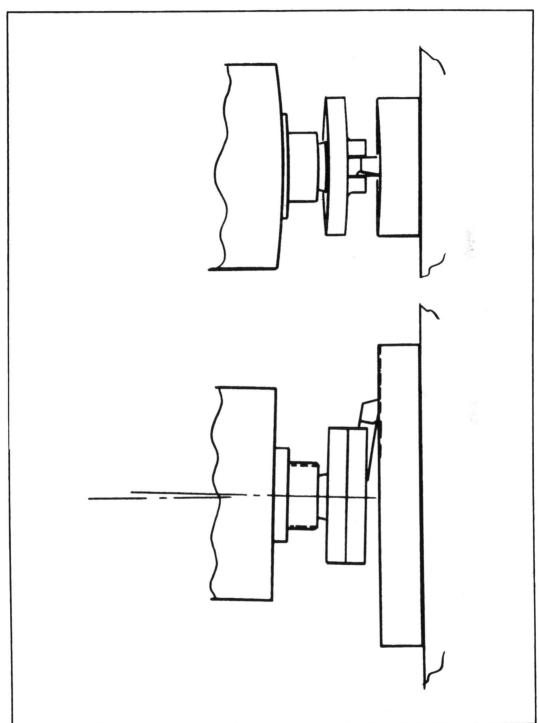

Fig. 7-18. The importance of squaring the milling head prior to using a fly cutter is illustrated here. Any tilt to the milling head will result in cutting a concave surface.

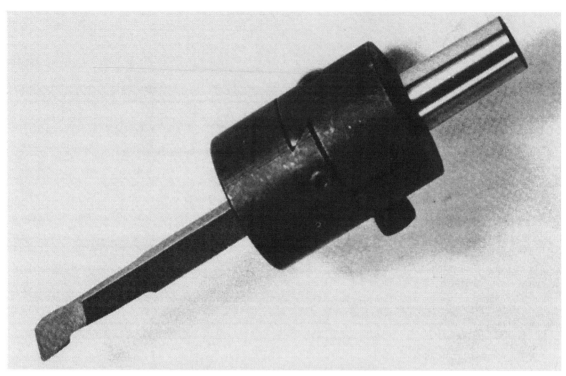

Fig. 7-19. Boring heads are used on milling machines to machine very accurate holes.

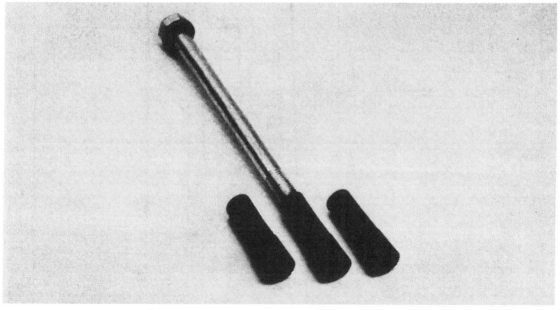

Fig. 7-20. Collet sets are used for mounting cutters in a milling machine.

144

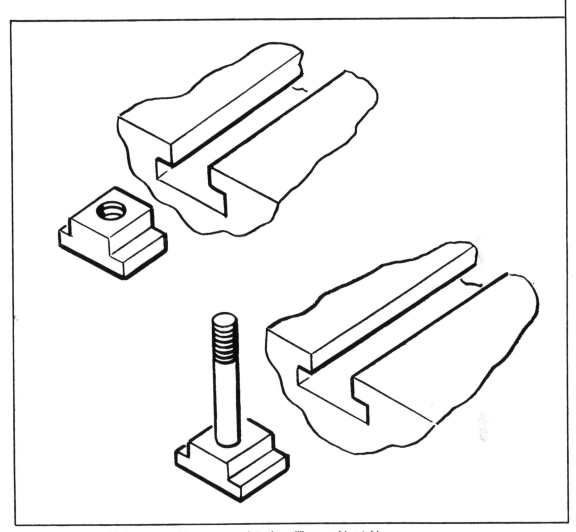

Fig. 7-21. T nuts and bolts are used to secure work to the milling machine table.

center of the clamp and threads into a T nut on the milling table. This bolt applies the pressure to clamp your work. When using one of these clamps, I like to place a small piece of scrap metal between the supporting bolt and the table. This spreads the force over a larger area and prevents damage to the table top.

Parallel Bars. Parallels (Fig. 7-23) are accurately machined spacer bars used to support irregularly shaped parts during various milling operations (Fig. 7-24). They can also be used to support parts away from the milling table to reduce the chance of accidentally cutting into the table top (Fig. 7-25). These bars are available from machine tool supply houses, but many machinists prefer to make their own, usually in matched pairs. All of the surfaces are machined so that they can be turned to give several different spacing heights. Sometimes a series of holes is bored into the bars to minimize the weight.

Angle Plates. Angle plates (Fig. 7-26) are used to make setups when the only machined sur-

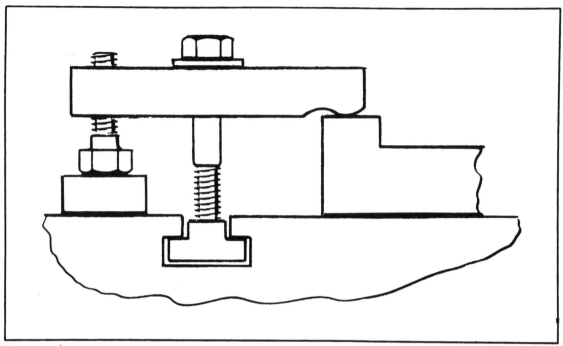

Fig. 7-22. Strap clamps are frequently used with the T nuts and bolts to secure work to the milling machine. A piece of scrap material under the head of the clamp leveling bolt helps to distribute the clamping force and prevents damage to the table top.

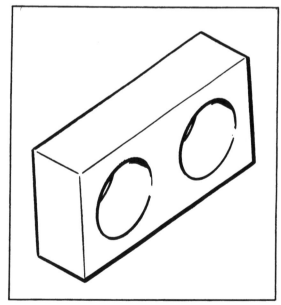

Fig. 7-23. Parallel bars are machined with all surfaces square to one another. They are quite useful in making setups for milling operations.

face on a part needs to be held in a vertical plane. As a rule these plates are fairly heavy, and they are gusseted to increase their rigidity. Angle plates are another example of tools that many machinists like to make for themselves (Fig. 7-27).

Sine Bars. It is sometimes necessary to make a setup where the part has to be rotated at an angle to the table. Sine bars (Fig. 7-28) are precision measuring devices that help the machinist make this type of setup. These bars are made with two round bars located on the bottom. The center-to-center spacing of the bars is held closely to either 5 or 10 inches. To establish any given angle, you use standard trigonometric table to find the sine of that angle. Then, if you have a 10-inch sine bar, place a spacer equal in height to the sine multiplied by 10 under one end of the sine bar. If you have a 5-inch sine bar, multiply the sine by 5. In either case the sine bar provides a very quick, accurate way to establish any angle. A sine bar is another example of a tool that many machinists like to make for them-

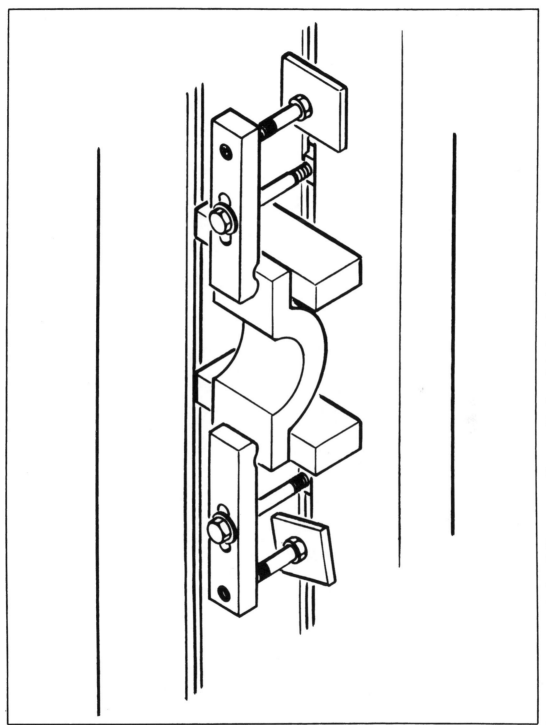

Fig. 7-24. Note how a pair of parallel bars are used to support an irregularly shaped part.

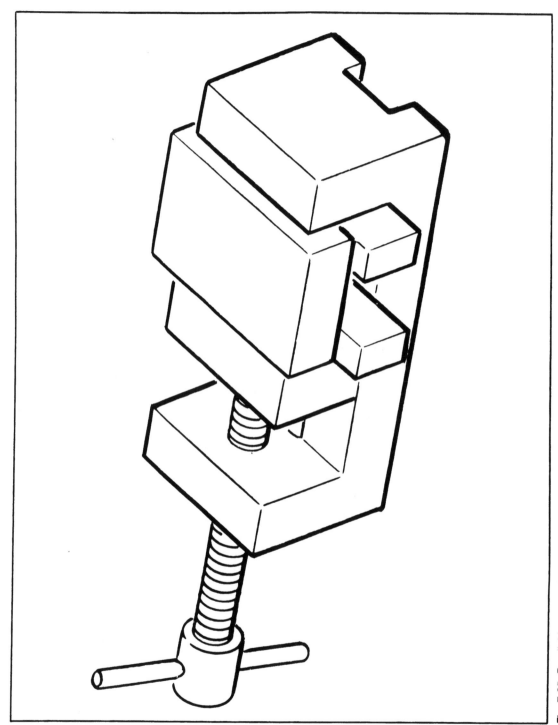

Fig. 7-25. Parallel bars can also be used to support a part when clamping it in a vise for machining.

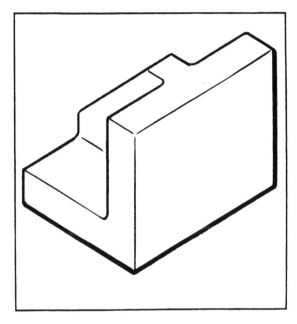

Fig. 7-26. Angle plates have two surfaces precision machined at 90° to each other.

selves. Sine tables may be found in any good, general book of trigonometry.

V Blocks. At times, it is necessary to hold a piece of round bar stock for milling a slot or a flat into it. V blocks (Fig. 7-29) provide an excellent way of holding any diameter round stock. If the bar is a short one, one V block will do the job. If the bar is very long, a matched pair of blocks should be used. Should you decide to make your own, make one long and then cut it in half so that you have a matched pair.

Milling Vise. A good milling vise (Fig. 7-30) is another accessory that should be purchased as soon as you obtain your milling machine. You will find that the vise can be used for most milling operations. The vise will simplify setups, reduce

Fig. 7-27. Angle plates are frequently used in setting up milling operations. An angle plate is used to support a bearing block prior to machining its base.

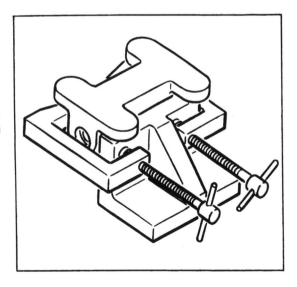

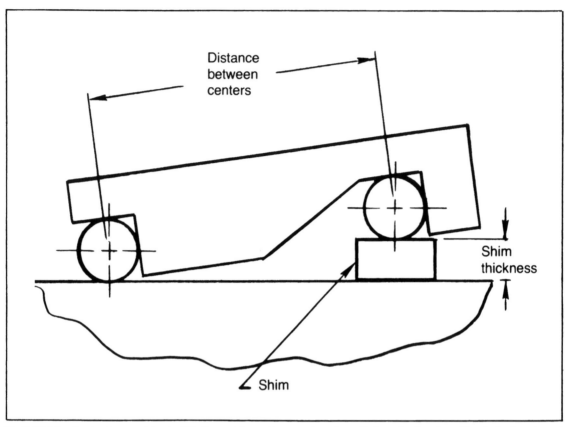

Fig. 7-28. The sine bar is a precision tool that is used for making angular setups. By knowing the sine of the desired angle and the distance between the two round bars on the sine bar, you can determine the exact shim thickness required to tilt the bar to the desired angle.

setup time, and enable you to hold small parts that would be difficult to hold any other way. A good milling vise should have provisions for bolting it to the table quickly and easily. It should also have the sides machined square to the fixed jaw to simplify setups. Still another nice feature to look for in a vise is a horizontal and a vertical V notch in the fixed jaw to be used for holding small diameter bar stock.

Center Finders. Center finders, sometimes called *wigglers*, are inexpensive accessories that really come in handy (Fig. 7-31). These clever devices are used to locate the center of the spindle in relation to a surface on your work. There are several types of center finders available, but they all work on the same principle (Fig. 7-32). The most common types consists of a spherical chuck and an indicator (Fig. 7-33).

A center finder is collet-mounted in the spindle in place of a cutter. The indicator is roughly aligned with the axis of the spindle so that it runs approximately true. Then, with the mill running, the free end of the indicator is brought in contact with the surface you are trying to locate. Because the indicator is not running perfectly true, it will only make intermittent contact with the part. As you move the part closer to the indicator, it will continue to make intermittent contact with the part. Each time it touches it will align itself a little closer to the axis of the spindle, so with each contact it will run a little truer.

150

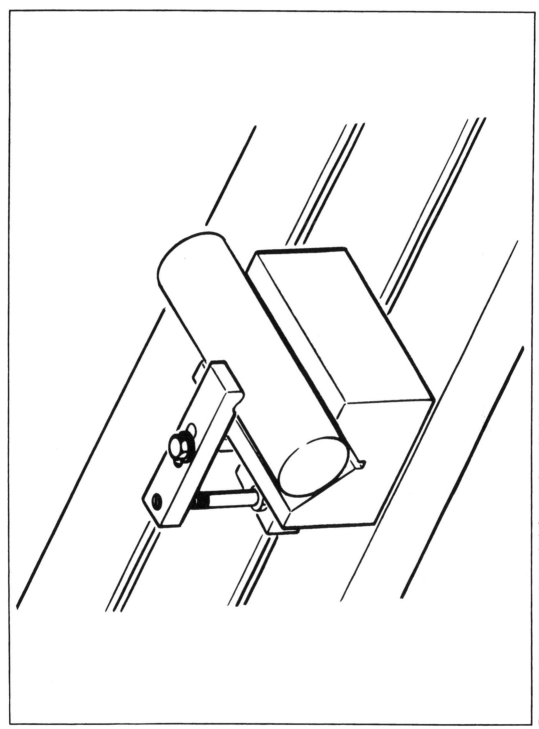

Fig. 7-29. V blocks are used extensively for centering and holding round bar stock. Used with a strap clamp, the V block provides an excellent way to hold round stock for any kind of a machining operation.

151

Fig. 7-30. Milling vises, such as this Sherline vise, are preferred for milling operations. By nature of their design, they clamp the part and apply a downward force that ensures more stability and greater holding power.

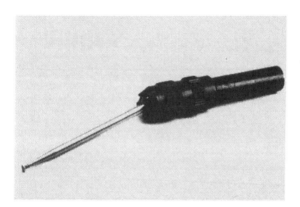

Fig. 7-31. Center finders or wigglers provide an easy method of locating a part in relation to the spindle on the milling machine.

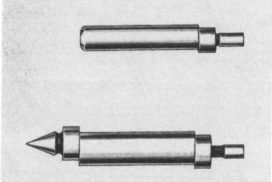

Fig. 7-32. Edge finders work on the same principle as the center finders (courtesy of L. S., Starrett Co., Athol, MA).

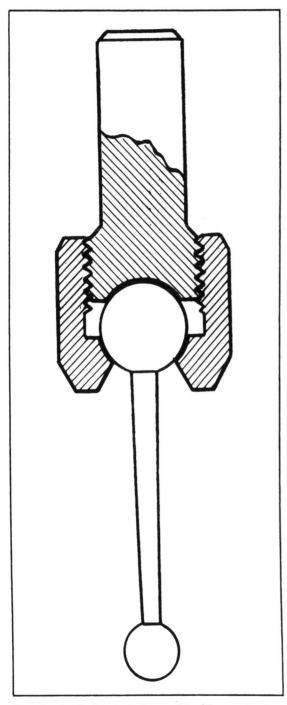

Fig. 7-33. A typical center finder consists of three parts: stem, clamp, and finder.

This process is continued until the indicator is running perfectly true. Then a strange thing happens. You don't have to guess whether the indicator is running true or not because the instant you go past that point, the indicator will swing over to the side. Now if you know the diameter of the indicator, it is a simple calculation to determine the exact distance from the center of the spindle to the edge of the part. Once you have this information, you can use the micrometer indicators on the X and Y lead screws to position the spindle anywhere you want it in relation to your part. This gives you the ability to locate holes or any other feature on any part with accuracy.

Dial Indicators. Whenever it is necessary to mount a part, angle plate, or vise on your milling machine, you will want to square the part to the table and the axis of the machine. There are several ways that this can be accomplished. For many parts, where a great deal of accuracy is not required, you can just eyeball it, or you can use a straightedge held firmly against the side of the bed or a T slot to position the part against. If you need to hold close tolerances, dial indicators (Fig. 7-34) provide the greatest accuracy.

A universal dial indicator graduated in 0.001-inch increments will do nicely. As a rule, the indicator is mounted on the spindle and held stationary as the work is moved back and forth by means of the milling machine's lead screws. The amount of misalignment is read directly on the dial indicator. The actual procedure for setting up a part is as follows:

1. The dial indicator is mounted on the spindle of the milling machine.

2. The part to be setup is mounted on the bed of the mill and secured in a vise with clamps or whatever hold-down devices seem appropriate. Hold-down bolts securing the vise or part to the milling table should be tightened just enough to hold firmly, but to allow some movement when the part is tapped with a small mallet.

3. The table and indicator next are positioned so that the indicator finger just comes in contact with the surface to be trued. If you are

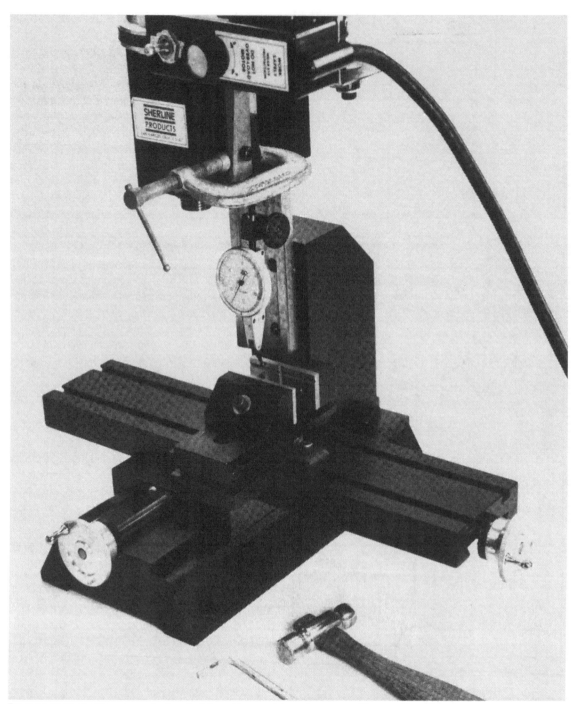

Fig. 7-34. Dial indicators are used extensively in making precision setups on a milling machine. This one is shown in use, setup for squaring a vise on a milling machine.

indicating on a vise, you should always indicate on the fixed jaw. Never indicate on the movable jaw.

4. Using the proper feedscrews, move the part towards the dial indicator finger to obtain a 0.005-inch to 0.010-inch deflection reading on the indicator.

5. Use the opposite axis feedscrew to move the part so that the surface being trued passes its entire length across the indicator finger. Any deflection of the indicator pointer indicates misalignment. Use your mallet to nudge the part into position. When you get the part located so that you have no further indication of misalignment, tighten the hold-down bolts. Check the final alignment after the hold-downs are tightened.

You can use a similar procedure to locate the spindle directly over a hole. Let the indicator finger touch the inside of the hole and while rotating the spindle by hand, observe the deflection on the indicator. Any deflection shown on the indicator shows a misalignment. You correct the position using the X and Y feedscrews and repeat the process until you see no more deflection on the indicator. You then are in perfect alignment with the hole. This same procedure can also be used to locate the spindle over the end of a vertical boss or shaft.

Chapter 8

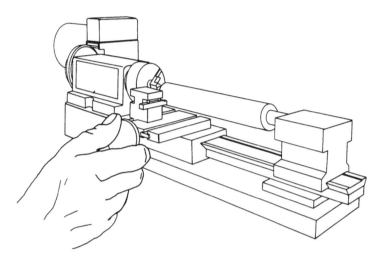

Milling Machine Operation

NOW THAT YOU HAVE BECOME FAMILIAR WITH the milling machine and its accessories, let's discuss the operation of one. I will begin by discussing a few basics, then you will be ready to start making chips.

UP AND DOWN MILLING

One of the most important things to learn is the difference between *up* (conventional) *milling* and *down* (climb) *milling*. In up milling, the cutter rotates opposite to the direction of feed. The cutting edge enters the part at the bottom of the cut and cuts upward until it clears the part. In down milling, the cutter is rotating in the direction of feed. The cutter enters the cut at the upper surface and cuts downward until it clears the part. See Fig. 8-1 for the difference between these two cuts. The difference is quite significant.

In up milling, the cutting force pushes the part away from the cut. If there is any play in the feedscrew, the cutting force pushes opposite to the direction of feed and prevents the play from causing

any problems. In down milling, the cutting force is pulling the part in the direction of feed. If there is any play in the feedscrew, the cutter may suddenly grab the part and pull it deeper into the cut. This could result in a broken cutter or a damaged part.

Another advantage of up milling is that the cutter enters into each cut more gradually. That is, the chip thickness increases as the cutting edge advances through the part. Because the increase in required cutting force is gradual, the shock of each cutting edge entering into the part is minimized. In down milling, just the opposite is true. The cutter encounters the maximum chip thickness at the start of each cut, so the shock is greater. As a result, there is more strain on both the cutter and the milling machine. In down milling, vibration is increased. There is more tendency for the cutter to flex away from the cut. Surface finish and dimensional accuracy are both affected.

For years, it has been standard practice in most machine shops to avoid down milling. Today, some commercial shops have learned to use down

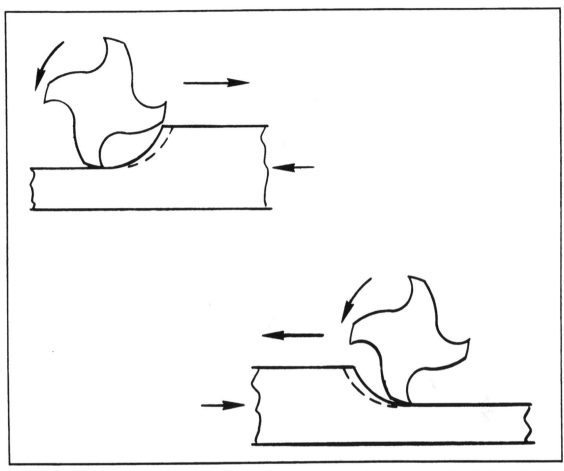

Fig. 8-1. In up milling (top), the work is fed into the cutter opposite to the cutter's direction of rotation. Note that the cutting force tends to push the work away and also that the depth of cut increases as the cutter progresses through the cut. In down milling (bottom), the work is fed in the same direction as the cutter's rotation. The cutting force will tend to pull the part into the cutter and could result in a broken cutter. Also, the depth of cut is at its deepest at the start of the cut. This will cause unwanted vibration and excessive strain on the milling machine.

milling to their advantage, but it is only used for special applications and on the best of equipment. *Down milling should never be attempted on any machine not equipped with an antibacklash mechanism.* For the type of work most of you will be doing, down milling should not be considered.

CONTROLLING DIMENSIONAL ACCURACY

The dimensional accuracy of a milling operation depends on several things. Proper clamping, feedscrew backlash, the rigidity of the cutter, and the rigidity of the part all play an important role. Let's look at each of them individually and see how they affect work.

Proper clamping is the first factor to consider (Fig. 8-2). Obviously, the work has to be secured to the milling table as rigidly as possible. There must be no chance of cutting force or vibration loosening or moving the part once it is secured in place. The first consideration in clamping a part should be to make sure all clamping surfaces are clean. This means the surfaces on the machine, as well as the

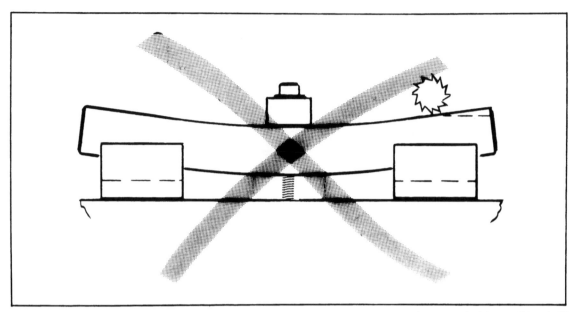

Fig. 8-2. An improperly clamped piece of bar stock may be distorted enough by the clamping force to make it impossible to hold the required tolerances.

surfaces on the part, should all be free of any and all foreign material like chips, dirt, or heavy rust.

After wiping the part and table with a clean shop cloth, most experienced machinists will make a final wipe with their bare hand. They are checking for burrs or other surface defects that might prevent the part from lying flat. Any high spots should be removed with a file.

The actual amount of clamping you need de-pends on the type of machining you are planning to do. Less clamping force is required for drilling and boring operations. More clamping force is required for milling operations. Also, less clamping force is required for taking light finishing cuts than would be required for heavy stock removal. In any case, the part should be clamped firmly, as close to the area being machined as possible.

Always take care that your clamps do not apply

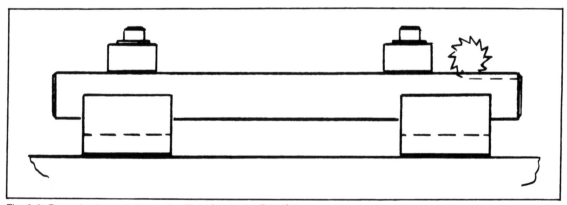

Fig. 8-3. Properly clamped, the work will not have any distortion.

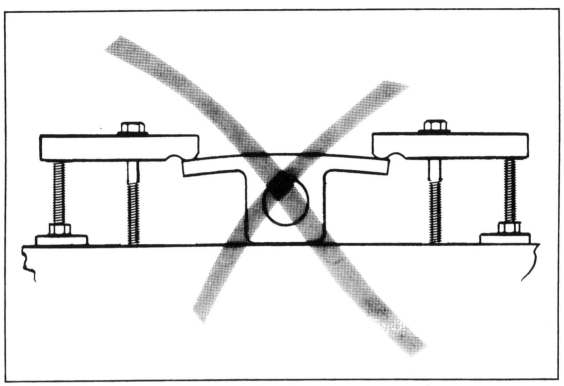

Fig. 8-4. Even parts that appear to be quite rigid may distort if improperly clamped.

pressure that will distort the part. As an example, it would not be advisable to take a long piece of bar stock, support it at both ends with V blocks, and then clamp it in the center with just one clamp. This might cause the bar to flex enough to throw your work out of tolerance (Fig. 8-2). It would be much better to use two clamps, one directly over each V block (Fig. 8-3). This same policy should carry

Fig. 8-5. The use of blocks or parallel bars to support the work directly under the clamp will prevent clamping distortion.

159

through in clamping any part. Irregularly shaped parts should, whenever possible, be solidly supported under each clamp.

Sometimes, especially on parts requiring extensive machining, it will be necessary to clamp a part in an area where the clamp will interfere with part of the milling operation. In these cases it may be necessary to use one or more extra clamps. Then each clamp can be removed temporarily to permit machining in a particular area.

In cases where you will be machining the side or periphery of a part, or where you are milling or drilling through a part, always mount the part on a piece of scrap material so that there will be no chance of the cutter running into the milling table.

The play in the feedscrews controlling the X, Y, and Z movement on a machine also contributes to inaccuracies in milling operations. All mechanical devices require some play or looseness to permit movement. The amount of play in a feedscrew is not critical as long as you remember that it is there. When you use the feedscrews to position a part for a cut, always move the part in a direction opposite to

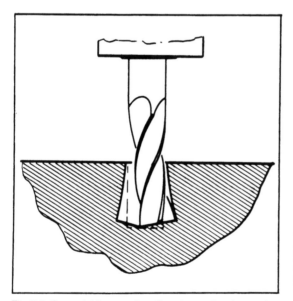

Fig. 8-6. Cutter rigidity can also affect the quality of a cut. Any flexing of the cutter in a plunge cut will result in a bell-shaped hole.

the direction of the cutting force. This will remove the play in the feedscrew.

The next consideration is the rigidity of the cutter (Fig. 8-6). The larger the diameter of the cutter, the more rigidity it will have. Within reason, it is best to use the largest diameter cutter that will do the job. I say within reason because with a small bench milling machine, you won't develop enough cutting force to get any appreciable deflection out of a ¼-inch diameter end mill. If you are using a large production-type machine, you might develop enough force to cause some deflection on a ½-inch diameter cutter.

Cutter mounting is also important. The cutter should always be inserted as deeply into the collet as possible. The amount of deflection you get from a cutter is very dependent on the ratio of diameter to unsupported length. Whenever it is necessary to use an extra long end mill or to have the cutter extend a long way out of the collet, take lighter cuts so that you minimize cutter deflection (Fig. 8-7).

Part rigidity is also a major consideration in controlling dimensional accuracy. Part rigidity is controlled by choice of material, wall thickness, and part configuration, and these are all controlled by the design and function of the part. Unlike cutter rigidity these things are usually beyond our control, but you can identify problem areas while you are making your setups. Once these areas have been identified, it's an easy job to reinforce them. Figure 8-8 shows how a piece of scrap can be clamped behind a slender wall to give it support. Similar methods can be used to reinforce other parts.

When these steps are not sufficient to eliminate tolerance problems caused by deflection, you still have one more variable left in your control. You can reduce the amount of force generated by the cutter; simply take lighter cuts or use slower feed rates. All finish cuts should be light cuts anyway. Heavy cuts should only be used for roughing operations.

Warpage is one last source of dimensional inaccuracies. Metals are quite stable for most applications. When you start working to tolerances of just a

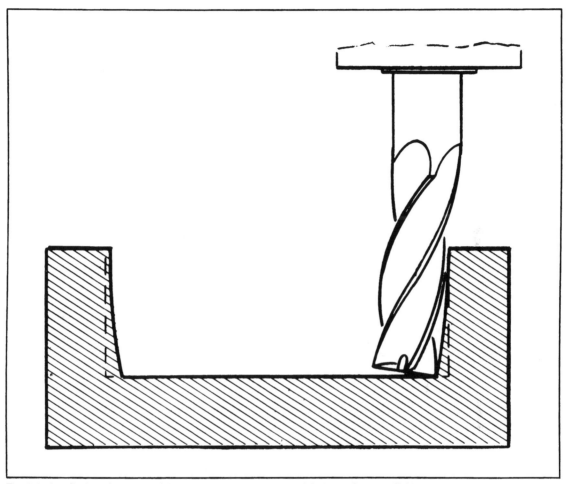

Fig. 8-7. Any flexing of a cutter while making deep cuts will result in inconsistent wall thickness.

few thousandths of an inch, though, you suddenly find that metals can be dimensionally unstable. For the most part these instabilities are caused by stresses built into the materials during their manufacture (Fig. 8-9). Unless the material is stress-relieved by heat treatment, these stresses can cause problems when you start machining. Because it is not always practical to stress-relieve before machining, it is necessary to follow machining practices that will minimize their effects.

Stresses are usually built up on the surface of materials that have been cold-roll finished. Many plate and sheet materials and some types of bar stock fall into this category. The best way to eliminate the effects of these stresses is to machine equal amounts of material from opposite surfaces. If you have a piece of ½-inch thick plate and you want to reduce it to ⅜ inch, don't machine ⅛ inch off of one side and expect the plate to remain flat. Instead, take 1/16 inch off of one side, then flip the plate over, clamp it flat, and machine off the remaining 1/16 inch. Always try to machine like thicknesses from each side of any material.

Stress can also occur in castings. They are the result of nonuniform cooling rates in the casting process. These stresses are more unpredictable

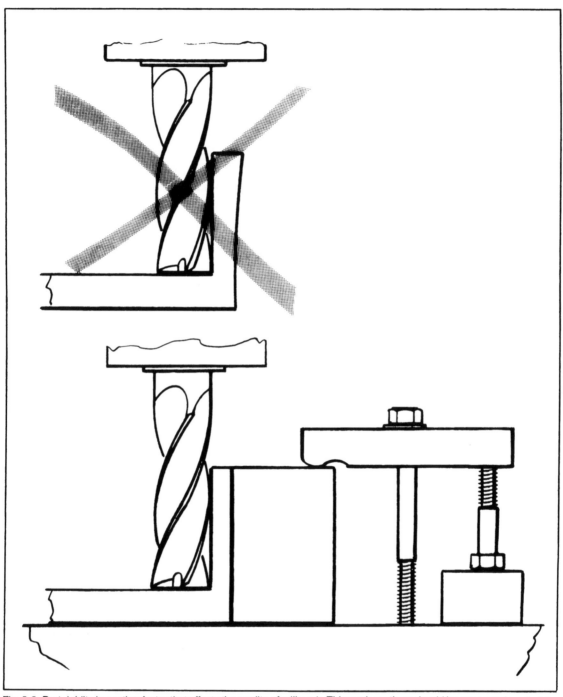

Fig. 8-8. Part rigidity is another factor that affects the quality of mill work. Thin work sections should be supported against the force of the cutter whenever possible.

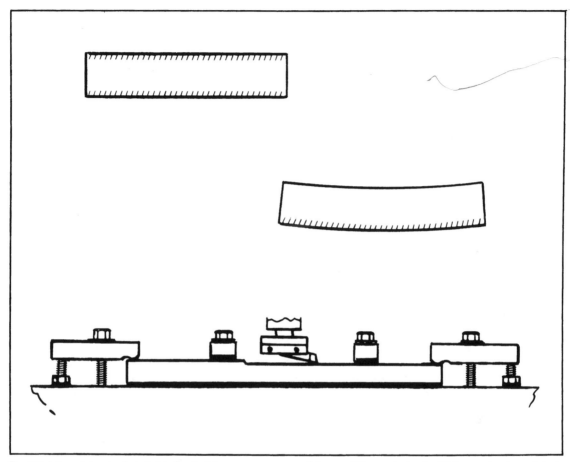

Fig. 8-9. Internal stresses in a piece of material may cause your work to warp. When reducing the thickness of a piece of stock, remove equal thicknesses from each side.

than those in plate or bar stock. The best way to compensate for them is to rough cut all machined surfaces prior to making any finish cuts.

SPEEDS AND FEEDS

As with all machining operations, cutter speeds and feed rates are quite important. The life of a cutter can be drastically reduced by either excessive cutter speeds or feed rates. Table 8-1 gives the maximum recommended spindle speeds for various size cutters cutting different materials. You will note that the chart shows maximum speeds only. Slower cutter speeds are acceptable; they only result in slower feed rates so that it takes a

little longer to complete a cut. On the other hand, excessive cutter speeds result in excessive strain on your equipment and overheats the cutter, which results in shortened cutter life.

You will also notice that in many cases the speeds shown for small diameter cutters are quite high. Many times these speeds may be greater than your milling machine is capable of turning. Just remember that as previously stated, slower cutter speeds are not a problem. If you can't achieve the r.p.m. shown on the chart, just use the fastest cutter speed you have available on your machine.

In most cases, cutter speed is related to the hardness of the material being cut. As a rule, the harder the material the slower the cutter speed. While this is true in most cases, there are exceptions. In Table 8-1 you will notice that there are two types of plastic listed. *Thermosetting plastics* are materials like phenolic, epoxy, and polyester. These materials are a resin in their uncured state and harden when exposed to heat. Once they have hardened, further exposure to heat will not soften them. For this reason you can use high cutter speeds when machining them. The heat generated by high cutter speeds will not affect them.

Most other plastics are *thermoforming*, and they soften when exposed to heat. The heat generated by high cutter speeds will cause these materials to soften and become sticky (Fig. 8-10). When this happens, the material begins to build up on the cutter. It doesn't take long for the material to completely fill up the flute area. When that happens, you have to remove the cutter and clean it. To eliminate this problem, use slower cutter speeds and use a cutting fluid to carry the heat away.

Plastics are only one good example of how excessive speeds can be detrimental in machining soft materials. There are other materials that are affected the same way. Some of these materials are lead alloys, copper, and soft aluminum. If you run into problems with any of these materials, the same techniques apply. Reducing cutter speed and using a suitable cutting fluid should make an appreciable difference.

Proper depth of cut and feed rates are a little more difficult to determine. They are dependent on factors such as cutter speed, the number of flutes, the type of material being cut, and whether or not you use a cutting fluid. Experience is the best teacher. Start your work with light cuts and slow feed rates. When you have completed a cut, examine the chips you have just produced. Discolored chips are evidence of excessive heat being generated. If the chips are not discolored and there was no apparent strain on the milling machine during the cut, you can try a heavier cut the next time.

In the small shop you are not concerned with high production rates. A few extra moments spent cutting a part slowly won't hurt. It is far better to be conservative and take a little longer than to risk damage to equipment by overstressing it. Remember that small diameter cutters break quite easily. The smaller the cutter, the lighter the cuts that must be taken.

Table 8-1. Recommended Maximum Cutter Speeds.

Material	Feet Per Minute	R.P.M. Based On Cutter Diameter							
		.062	.125	.187	.250	.375	.500	.750	1.000
Wood	600	37000	18500	12300	9250	6150	4625	3075	2312
Brass	600	37000	18500	12300	9250	6150	4625	3075	2312
Plastic Thermosets	400	24600	12300	8215	6150	4107	3075	2053	1532
Aluminum Extrusions	400	24600	12300	8215	6150	4107	3075	2053	1532
Aluminum Plate	400	24600	12300	8215	6150	4107	3075	2053	1532
Bronze	250	15400	7700	5133	3850	2566	1925	1283	962
Aluminum Cast	200	12300	6150	4107	3075	2053	1533	1026	716
Copper	120	7400	3700	2466	1850	1233	925	616	462
Cast Iron	100	6150	3075	2050	1537	1025	768	512	384
Steel, Low-Carbon	80	4930	2465	1643	1232	821	616	410	308
Steel, High-Carbon	70	4930	2150	1433	1075	716	537	358	268
Tool Steel	60	3700	1850	1233	925	616	462	308	231
Stainless Steel									
Free Cutting	60	3700	1850	1233	925	616	462	308	231
All Others	30	1850	925	616	462	308	231	154	115
Plastic Thermoformed	50	3075	1537	1025	718	512	384	256	192

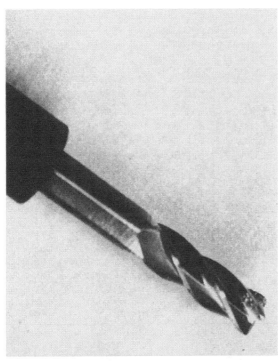

Fig. 8-10. Heat generated by high cutter speeds will cause some materials to become soft and sticky. As a result, the material will build up on the cutter and eventually fill the flutes. Slower cutter speeds and cutting fluid will help to eliminate this problem.

FINISHES

The smoothness of the finish on a milled surface is dependent on three things: the sharpness of the cutter, the number of flutes on the cutter, and the feed rate (Fig. 8-11). A dull cutter will have a tendency to tear the material it is cutting. Consequently, even with a very slow feed rate, the resulting finish will be somewhat rough. For this reason you should use your older and duller cutters for roughing work. Save your sharp cutters for finish cuts.

The more teeth a cutter has, the smoother the cut will be for any given feed rate (Fig. 8-12). Even a single point tool will produce a smooth surface if fed slowly. The secret is to keep the cutter moving across the part at a uniform rate. An uneven feed rate results not only in uneven surface texture on the part, but also causes some deviation in the resulting depth of cut.

When cutting a small pocket into a part, the pocket is cut by making a series of cuts, each one a little deeper than the preceding one. This is continued until the desired depth of cut is achieved. The resulting side wall will show the marks of each succeeding plunge. To make the pocket look more attractive and professional, the pocket should be roughed out slightly undersize. Then a light finish cut can be taken the full depth of the pocket, or at least as deep as the flute length on the cutter. The resulting pocket will be much cleaner looking.

Engine turning is one method of producing an attractive finish on a part (Fig. 8-13). This method was used extensively years ago, but because it is a slow process it is not seen much anymore. The finish is produced by using a wooden dowel and some fine grinding or lapping compound. Mount the dowel in your milling machine in place of a cutter, then apply a little lapping compound to the end of the dowel. Next, at regular intervals of approximately half the diameter of the dowel, bring the dowel in contact with your part long enough to produce a swirling pattern on the part (Fig. 8-14). When one row of swirls is complete, index the part over again approximately half the diameter of the dowel and start a new row of swirls. The procedure is repeated over and over until the entire surface of the part is covered with overlapping swirl patterns. (Other methods of finishing a part are discussed in Chapter 10.)

See Figs. 8-15 and 8-16 for examples of chamfering and key slot cutting that can be machined in the small shop.

SAFETY

Whether you are using a large machine or a small one, you can't put too much emphasis on safety. The same safety rules apply to milling machines as to other power tools, plus a few more. Take time to read them and be sure to follow them.

● Never wear loose clothing when operating a milling machine.

Fig. 8-11. Notice the finish left by this two-flute cutter.

Fig. 8-12. This is the same part shown in Fig. 8-11. Notice the difference in finish using a four-flute cutter. The feed rates were approximately the same.

Fig. 8-13. A wooden dowel and grinding compound are used to produce an engine turned finish.

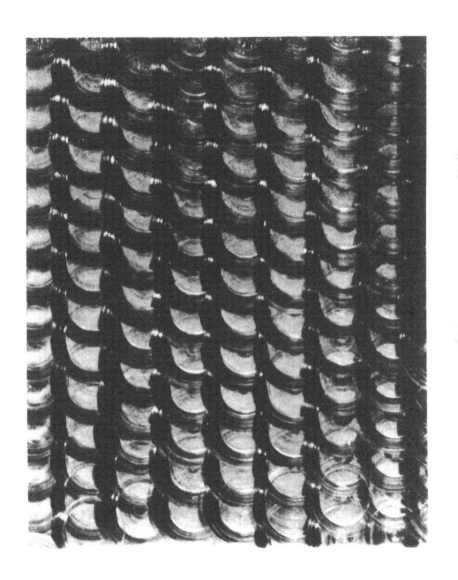

Fig. 8-14. A finished engine turned surface.

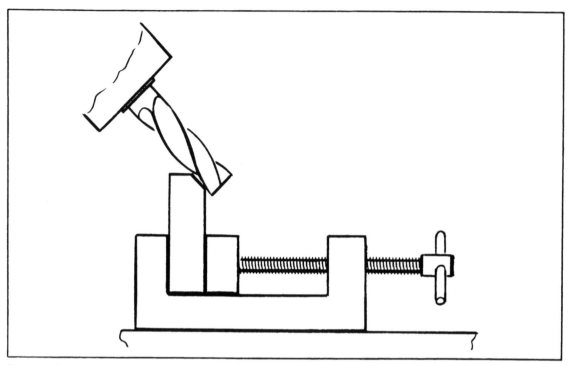

Fig. 8-15. Small chamfers are machined on a part by rotating the milling head to the desired angle.

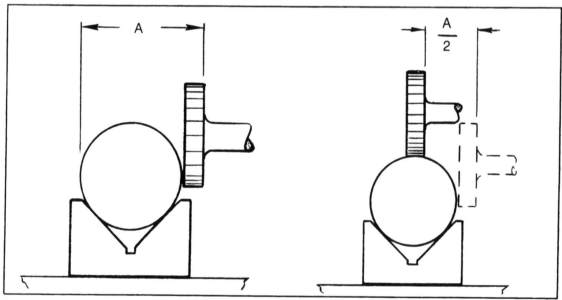

Fig. 8-16. To locate a key slotting cutter to the center of a shaft, bring the cutter into contact with the side of the shaft. Then, using the micrometer dials on the feedscrews, move the cutter over exactly half the sum of the shaft diameter and the cutter width.

● Remove all jewelry before operating a milling machine. This includes rings, watches, idenfitication bracelets, etc.

● Always wear safety glasses.

● Keep your hands clear of moving parts. Remember that the side cutting edges of an end mill are much sharper than the side cutting edges of a drill. Use caution when handling these tools.

● Never remove chips with your bare hands, especially when the cutter is turning. Always use a brush or an air hose.

● Never try to take measurements while the machine is running.

● Use common sense.

Chapter 9

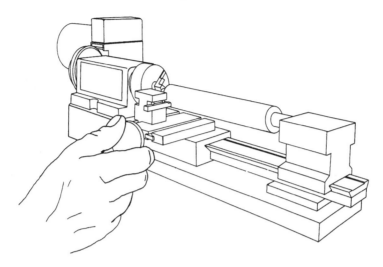

Material Selection

MOST OF US ARE GUILTY OF LOOKING AT AN object and saying "that's made out of aluminum" or "that's made out of plastic." We tend to oversimplify material identification. There are many kinds of plastic, aluminum alloys, and steel alloys, each with its own set of properties and peculiarities. As a machinist you should be familiar with some of the more common materials, their properties, and their uses.

The materials that most of you will be working with can be broken down into three basic groups: ferrous metals, nonferrous metals, and plastics. *Ferrous* materials are important because they include the strongest and hardest materials with which most of you will be working, and also because they are easily heat-treated to change or modify their mechanical properties. You can soften ferrous materials to facilitate machining, and you can also harden them to make them stronger, harder, and more resistant to wear.

The nonferrous group of metals includes aluminum alloys, copper alloys, and lead. These materials are important for many reasons. Alu-minum alloys are lightweight and easy to machine. Copper alloys, like brass, are important because of their decorative properties, their ease of machining, and also because they make excellent bearing materials. Lead is important because of its weight, low melting temperature, and softness.

Like steel (a ferrous metal), many of these nonferrous materials can be heat-treated, but generally the processes are too complicated to be accomplished in a small workshop. If you want a nonferrous material in a specific heat-treated condition, purchase it as raw stock, preheat-treated to that specific condition. Many nonferrous materials are soft and easy to machine compared to steel, even when they are in a hardened condition. In some cases they are actually easier to machine in the hardened condition than they would be in a soft condition, because the hardened materials are less gummy and there is less tendency for them to stick to and build up on the cutting tool.

Certain plastics are also important because of their decorative properties and also because some make an excellent bearing material. Of the many types of plastic available, I will only discuss four specific types: acrylic, polycarbonate, nylon, and polytetrafluoroethylene.

MECHANICAL PROPERTIES OF MATERIALS

There are many terms used to describe the properties of materials (Fig. 9-1). The following are only a few of the more common ones.

Strength. The strength of a material is the measure of its ability to resist deformation under a steady or continuous load. The strength of a material can be measured and expressed in four different ways: tensile strength, compressive strength, shear strength, and torsional strength. Materials that rate high in one kind of strength test may not rate high in another. For this reason, if strength is a critical consideration in selecting a material, consider how the part will be used before deciding which strength rating is the most meaningful to you.

Strength is always expressed in pounds per square inch (psi). The most commonly used expression of strength is *tensile strength* (Fig. 9-2). The tensile strength of a material is that amount of force, applied as tension, required to fracture a sample of that material having a 1-square-inch cross section. When you know the strength of a material, you can calculate the strength of any part made from that material. Simply multiply the smallest cross section of the part times the strength of the material. For example, to find the tensional strength of a part having a cross-sectional area of ½ square inch and made from a material with a tensile strength of 100,000 psi, simply multiply:

$$100,000 \times .5 = 50,000 \text{ pounds}$$

A term frequently used with tensile strength is *yield point*. Usually, a part in tension will start to stretch or elongate before it fractures. The yield point is the amount of force required to permanently stretch or elongate a part. As a rule, the yield point is just slightly lower than the tensile strength.

Hardness. Hardness describes a material's resistance to denting, scratching, or permanent deformation. Hardness can be varied in some materials by the process of heat treating. Other materials increase in hardness only as a result of being cold-worked. When a material hardens as a result of cold working, it is said to be *work-hardened. Cold working* is the process of mechanically deforming a material either by hammering or rolling when the material is at room temperature.

Hardenability. Hardenability is a relative term that describes a material's ability to respond to heat treating. Materials are usually rated as good, moderate, or poor in hardenability.

Brittleness. Brittleness describes a material's inability to be permanently deformed. Cast iron is a good example of a brittle material. Under a heavy load or sudden impact, it will fracture rather than bend or deform. In most metals brittleness, hardness, and strength are all closely related. As an example, when you heat-treat steel to increase its hardness, you also increase its strength and brittleness.

Malleability. Malleability is the opposite of brittleness. It is the ability of a material to be permanently deformed by compression without fracturing.

Ductility. Ductility is the ability of a metal to be drawn through a die to change its shape or size (Fig. 9-5). The tensile strength of a ductile material must be high enough that the material will assume the shape of the die rather than break. All ductile materials are malleable, but not all malleable materials are ductile.

Toughness. Toughness is a material's ability to resist shock or sudden impact without fracturing.

Machinability. Machinability is the ease with which a material can be machined. It deals primarily with the maximum rate that metal can be removed, the quality of the finished surface, and the life of the cutting tool. Machinability is generally stated in comparison to the machinability of a standard. For steel, the standard is SAE alloy 1112, for aluminum it is alloy 2011, and for copper alloys the standard is alloy 360. Sometimes machinability is stated loosely as poor, fair, good, or excellent.

174

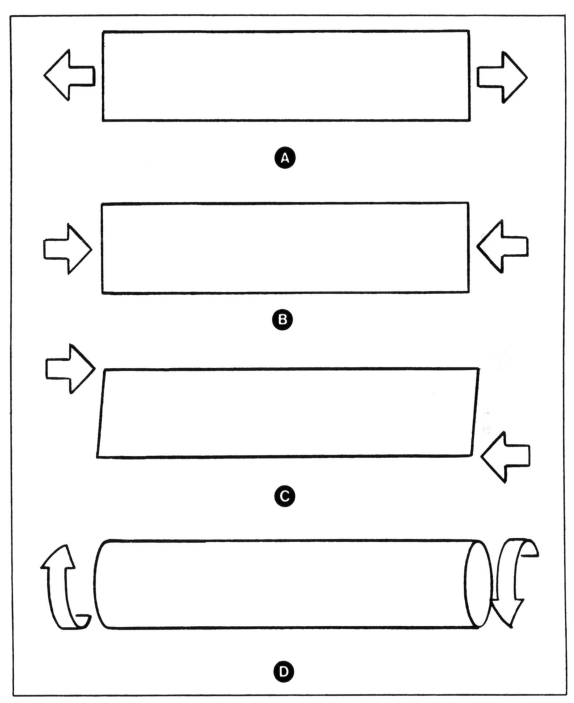

Fig. 9-1. The strength of a material is usually expressed in one of four ways: (A) tensile strength, (B) compressive strength, (C) shear strength, and (D) torsional strength.

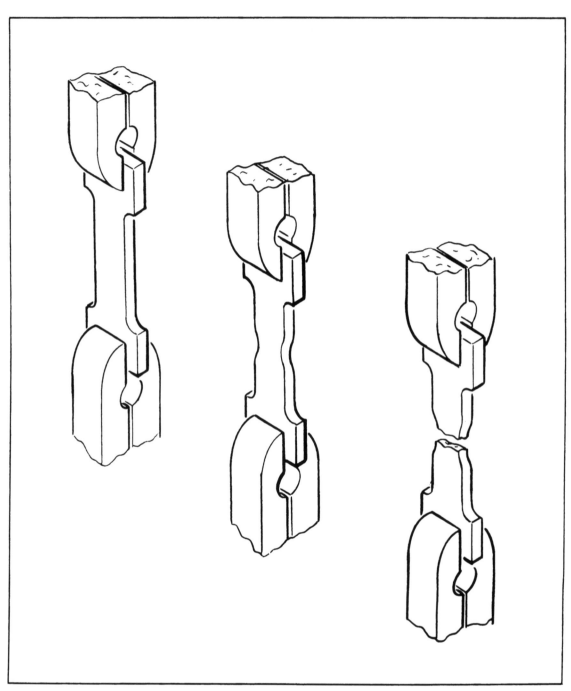

Fig. 9-2. When a material is tested for tensile strength, a sample with a known cross-sectional area is subjected to increasing tensile loads. The force required to actually start stretching the sample is called the yield point. The force required to fracture the sample is the tensile strength.

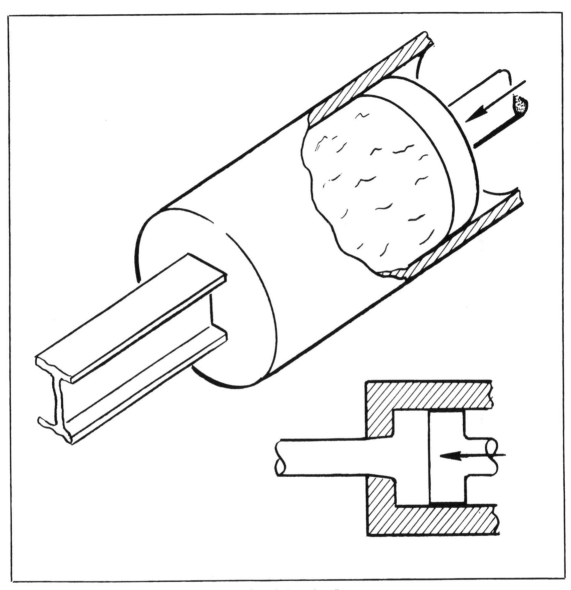

Fig. 9-3. Metal is extruded by applying pressure to force it through a die.

Otherwise it is expressed in percentage terms with the standard rated at 100 percent.

HOT-ROLLED AND COLD-ROLLED MATERIALS

Hot-rolled and *cold-rolled* (Fig. 9-6) are two terms that are frequently heard when discussing metals. The terms actually describe differences in

the manufacturing process, not the material itself, but the two different manufacturing processes have a big influence on the final properties of the material.

In the manufacturing of sheet and plate metals the material usually starts out as a cast billet or ingot. These billets are then passed through a

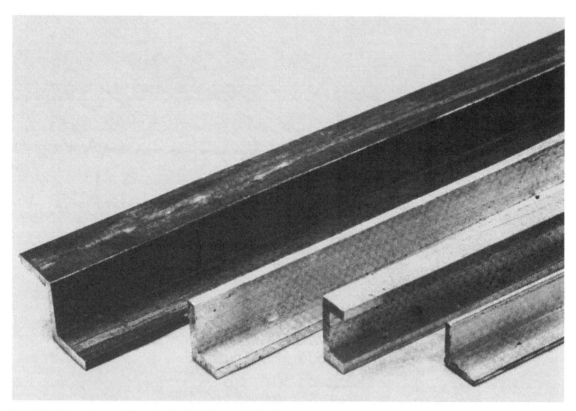

Fig. 9-4. A few of many different extruded shapes of aluminum.

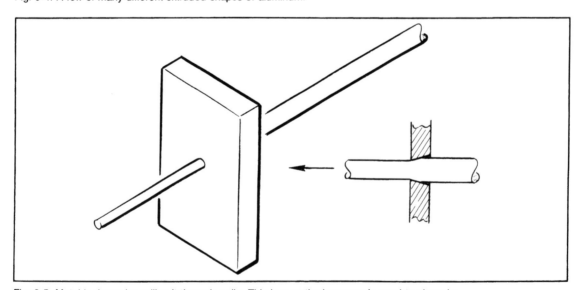

Fig. 9-5. Metal is drawn by pulling it through a die. This is a method commonly used to size wire.

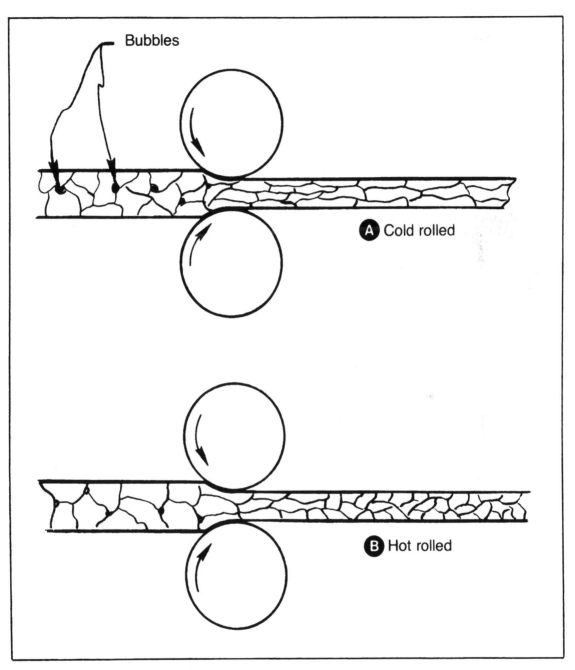

Fig. 9-6. Rolling a material compresses it, eliminating small bubbles and discontinuities inherent in cast materials. Cold rolling is accomplished with the material at room temperature. The rolling process causes the grain of the material to be severely deformed and causes internal stresses. These stresses may cause warpage problems for the machinist. Hot rolling is accomplished with the material heated sufficiently to permit reformation of the grain after rolling. The new grains are generally smaller and finer than in the original cast material resulting in a stronger, denser material.

series of rolling presses that form the material to the desired thickness. The rolling process can be done with the material heated to a red heat so that it is soft and easily formed, or it can be done cold, with the material at room temperature. There are significant differences in the resulting mechanical properties. These differences apply to both ferrous and nonferrous metals alike.

The process of rolling the material to size after it has been cast actually improves its quality. Grain size is reduced, porosity is eliminated, and mechanical discontinuities are eliminated. The resulting material is denser and exhibits improved mechanical characteristics.

Hot rolling is done at temperatures high enough to permit complete reformation of the material's grain after rolling. The new grains are generally smaller and finer than the original cast grain, which results in a stronger denser material. Also, the growth of the new grain eliminates stresses within the resulting sheet or plate. This reduces the material's tendency to warp when it is machined. Hot-rolled steel can usually be identified by its dark, somewhat scaly surfaces. See Figs. 9-7 through 9-14 for an example of how steel is processed.

Cold rolling is usually accomplished with the material at room temperature. In this process the grain of the material is distorted into fibers that run in the direction of rolling. The fibers account for the fact that cold-rolled materials are generally stronger in one axis than they are in the other. With most materials, cold rolling work-hardens it to increase strength, hardness, and brittleness, all in proportion to the severity of deformation. In cold rolling, a material's surface grains are generally deformed more severely than the inner grains. As a result, the material has internal stresses that may cause warpage during machining.

Cold-rolled materials can be recrystallized by reheating. This process is called *annealing* or *stress relieving*. When the material is heated enough to permit the grain structure to reform, all of the material's properties revert back to their original status.

FERROUS METALS

Ferrous metals are iron or materials that contain iron, including stainless steels. (In this book I will discuss stainless steel separately.)

Pure iron has few industrial uses and is seldom seen in the machine shop. It is a soft, ductile metal, not hardenable by heat treatment, and only slightly hardenable by cold working. It is highly magnetic in the presence of a magnetic field but will not retain its magnetism. It is also highly susceptible to rusting. The properties of iron change considerably when it is alloyed or mixed with small amounts of other elements.

Wrought Iron. Wrought iron is probably the oldest form of iron used by man. It is a mixture of iron and impurities such as sulfur, phosphorus, manganese, and silicon. These impurities are found in iron ore, and in olden times they were not completely removed during the primitive refining processes. The amounts of these impurities in primitive wrought iron varied from one batch to the next so that the properties of this material were not very consistent.

Today, using modern refining methods, the amounts of these impurities are carefully controlled to exacting standards. Today's wrought iron is a tough, malleable, ductile material. It is nonhardenable and fairly resistant to corrosion. It welds easily but is difficult to machine. It is typically used in decorative wrought iron products such as fences, gates, and gas and water pipes. It is also used as a base metal for galvanized sheet iron.

Steel. Steel is iron that has been alloyed with carbon and other alloying elements to alter its properties. Some of the more common alloying elements are chrome, manganese, nickel, phosphorus, sulfur, and lead. The Society of Automotive Engineers (SAE) has established a standard numbering system for specifying various steel alloys. This system uses a four digit number, with the first

Fig. 9-7. Blast Furnace L, one of the largest in the Western Hemisphere, a 300-foot high giant (courtesy of Bethlehem Steel Corp., Sparrows Point, MD).

Fig. 9-8. Molten pig iron from a blast furnace is poured into an open hearth furnace over previously added limestone, iron ore, and steel scrap where it is refined and purified into steel at temperatures of approximately 2900° F. (courtesy of Bethlehem Steel Corp., Sparrows Point, MD).

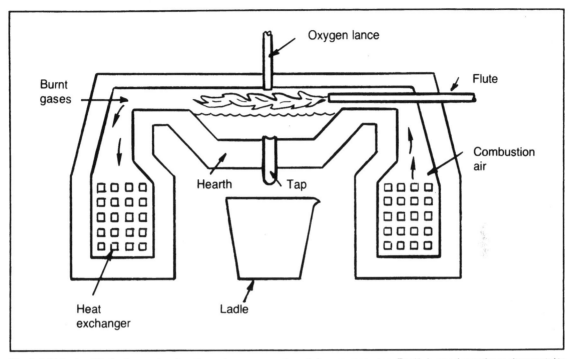

Fig. 9-9. This simplified cutaway drawing shows how an open-hearth furnace operates. Fuel is burned over the molten metal to produce the heat. The burnt gases are then exhausted through a heat exchanger to preheat the incoming combustion air. Oxygen injected into the system from an overhead oxygen lance speeds the refining process. Furnace capacities range from 100 to 500 tons.

digit identifying the principal alloying element other than carbon. The second digit identifies the other alloying elements, and the last two digits specify the carbon content in points. One hundred points are equal to 1 percent. Table 9-1 identifies the more common alloys and indicates basic properties.

Plain Carbon Steel. Carbon is the principal alloying element added to iron to make steel. The amount of carbon combined with the iron is usually specified in points or percentage of the total weight. The more carbon added, the stronger the resulting steel will be. The maximum amount of carbon used seldom exceeds 1.7 percent or 170 points. You can quickly calculate the approximate tensile strength of annealed low carbon and medium carbon steels by simply adding 1000 pounds per square inch (psi) to the tensile strength of iron (40,000 psi) for each point of carbon.

Example: The approximate tensile strength of SAE 1018 steel (plain carbon steel with an 18 point carbon content) is:

$$40,000 + (18 \times 1000) = 58,000 \text{ psi}$$

This formula holds approximately true for any plain carbon steel with a carbon content up to 80 points. The rate of increase in tensile strength falls off rapidly after 80 points.

The carbon content affects the strength and hardenability of steel. The greater the carbon content, the greater the degree of hardenability. Plain carbon steels are usually classified in one of three general categories: low-carbon steel, medium-carbon steel, or high-carbon steel.

Low-carbon steel, also called *mild steel*, is any steel with a carbon content less than 0.2 percent. These steels are good structural steels. They are

183

Fig. 9-10. A dramatic spectacle: charging molten iron into a basic oxygen furnace. There furnaces produce high-quality steel much faster than the open-hearth furnace (courtesy of Bethlehem Steel Corp., Sparrows Point, MD).

Fig. 9-11. Molten iron rushes from a blast furnace through a series of clay-lined runners into a "submarine" (a special railroad car). The iron will be transported to a different location where it will be charged, still in a molten state, into a basic oxygen furnace for refinement into steel (courtesy of Bethlehem Steel Corp., Sparrows Point, MD).

185

Fig. 9-12. This is a segment of the 85-foot long control panel of a huge new computer-controlled blast furnace (courtesy of Bethlehem Steel Corp., Sparrows Point, MD).

Fig. 9-13. Molten steel produced in electric furnaces is poured from a giant ladle into ingot molds. The ingots will eventually be processed through a rolling mill into plate, sheet, or structural shapes (courtesy of Bethlehem Steel Corp., Sparrows Point, MD).

Fig. 9-14. Molten steel is cast into one long continuous slab, which is then flame cut into smaller slabs. These slabs are then rolled to form sheet or plate stock (courtesy of Kaiser Steel Corp., Fontana, CA).

Table 9-1. Basic S.A.E. Numbering System for Steel Alloys.

Type of Steel	Number	Uses And Characteristics
Carbon Steels		
Plain Carbon	10XX	General purpose
Resulfurized Plain Carbon	11XX	Free machining, general purpose
Rephosphorized Plain Carbon	12XX	Free machining, general purpose
Manganese Steels	13XX	Tough. Good shock and abrasion resistance.
Nickel Steels		
3.5% Nickel	23XX	Tough. Excellent abrasion resistance.
5% Nickel	25XX	Tough. Excellent abrasion resistance.
Nickel-Chromium Steel		
1.25% NI, .65% Cr	31XX	Very hard, Strong and tough.
3.5 %, 1.57% Cr	33XX	Used for gears, shafts and springs.
Molybdenum Steels	40XX	High strength structural steel
Chromium-Molybdenum Steels	41XX	retains strength at elevated temperatures. Used for bearings, automobile parts, and machine parts.
Chromium Steel	50XX	Hard and tough. Used for springs, bearings, cutting tools, Etc.
Chromium, Vanadium	60XX	Very hard and strong. Used for auto parts, gears, springs, axles, and tools.

Note: S.A.E. numbers are basically the same as the A.I.S.I. (American Iron and Steel Institute) numbers. A.I.S.I. numbers sometimes prefix with a letter which identifies the type of refining process used to produce the steel. The letter L between the second and third digit indicates the addition of lead to improve the machinability rating of the steel.

ductile, malleable, and weldable, but rated only fair in machinability. They are considered nonhardenable, but they can be case-hardened. *Case hardening* is a process that increases the carbon content of the surface of a part. (The process is discussed in more detail in Chapter 10.) Typical uses for low-carbon steels include nails, nuts, bolts, screws, washers, wire, structural parts, automobile frames, body parts, and so on. Low-carbon steel is available in bars, sheets, plates, and various structural shapes.

Medium-carbon steels contain from .2 percent to .6 percent carbon. They are both harder and stronger than low-carbon steels, and they can be heat-treated. These steels do not weld well with oxyacetylene, but they can be brazed or arc welded. Machinability is rated as fair. Medium-carbon steels find wide application in machine parts such as gears, shafts, axles, and mechanical linkages. These steels are available in rod or bar stock and in sheet and strips.

High-carbon steels may contain anywhere from .6 percent to 1.7 percent carbon. Carbon levels above 1.3 percent, however are not common. These materials are used wherever a high degree of strength, hardness, or resistance to wear is required. Because of their high strength, these materials are generally difficult to machine. Parts made from high-carbon steel are usually heat-treated to optimize their mechanical properties.

Typical applications for high-carbon steels include many tools such as drills, taps, files, saws, hammers, chisels, etc. They also include farm tools such as shovel blades, plow shares, scrapers, and disks. High-carbon steels are also used extensively to make springs. These materials are available both as hot and cold-rolled steel in bars, rods, and plates. In many cases high-carbon steel may be purchased premachined and ground to closely toleranced dimensions.

Free-Machining Carbon Steel. The addition of sulfur or phosphorus to plain carbon steel improves its machinability rating considerably. As an example, SAE 1112 is a resulfurized steel, rated at 100 percent. Its plain carbon steel equivalent is SAE 1012, which is rated at 53 percent, only about half as good as SAE 1112. In addition to the improvement in machinability, resulfurized and rephosphorized steels are somewhat stronger and harder than their plain carbon equivalents. Other properties of resulfurized steel are about the same as for plain carbon steels, with the exception of welding. The addition of sulfur generally makes the steel more difficult to weld.

Free-machining carbon steels are used extensively whenever machinability is important. These steels will be the preferred material for many of the jobs tackled in small machine shops. Cutting speeds may be increased up to four times over the speeds used for plain carbon steel, and the resulting finishes are excellent.

Alloy Steels. Alloy steels are those that contain one or more alloying elements (elements other than iron and carbon). There are more than 25 different elements that are used in various combinations to make alloy steels.

Nickel. Nickel added to steel in small amounts (.3 to 3.7 percent) increases the strength, toughness, and resistance to abrasion and corrosion. It is frequently used with chrome, and the resulting alloys make an excellent forging material. Nickel-chrome steels are frequently used to make structural machine parts subject to severe working conditions. Nickel used in large quantities up to 22 percent is a principal ingredient in stainless steel.

Chromium. Chromium, alloyed in small amounts (.3 to 1.6 percent), either by itself or in conjunction with other elements, increases toughness, hardenability, resistance to abrasion, and resistance to corrosion. These alloys are much stronger than plain carbon steel with proper heat treating. Typical uses include bearing races, bearing balls, gears, shafting, and springs. Used in large amounts, 11 to 26 percent, chrome is a principal ingredient of stainless steel. Chrome is also a principal ingredient in high speed tool steels.

Molybdenum. Molybdenum is frequently used to increase the hardenability of steel. It also improves toughness and resistance to shock. Molybdenum is frequently used with chromium to make chrome-moly steels such as SAE 4130. This material can be machined to a good finish, has good resistance to corrosion, and in many cases can be hardened to near tool steel properties.

Cobalt and Tungsten. These elements are principal ingredients in tool steels. Tungsten is used in amounts varying from 2 to 20 percent. It produces a fine grain structure that enables cutting tools to retain their sharpness. It also improves the hardenability of steel. Cobalt is used in amounts varying from 5 to 55 percent. It is used primarily to improve high temperature hardness.

Lead. Lead is sometimes added to steel to help improve machinability. In small quantities it has no effect on the strength of the steel, but it does make machining a lot easier. When steel has been leaded, it is indicated in the alloy designation number by adding a capital L between the second and third digit of the number.

STEEL ALLOY IDENTIFICATION

The average small shop machinist may find himself working with bits and pieces of steel gleaned from scrap piles somewhere. This means that most of the steel he is working with will be unidentified. Short of chemical analysis, there are no good accurate methods of identifying the steel, but sometimes you can make good guesses. You can make some judgment based on the color and finish of the material, and you can use a file to determine the hardness. The best method of identification is the *spark test*.

All ferrous metals produce sparks when ground on a grinding wheel. Fortunately, different types of steel produce different kinds of sparks. With some practice, anyone can learn to read the spark patterns and thereby make identification. Figure 9-15 can be used to help identify the various spark patterns. Even better, you can compare a few

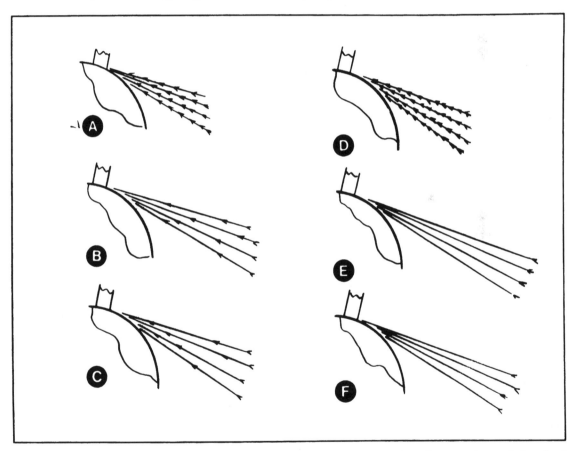

Fig. 9-15. Steel alloys can be identified by their appearance and by the sparks they produce when ground on a grinding wheel. (A) Cast iron—Gray in color. It produces short, light orange sparks with lots of spurs in close to the grinding wheel. (B) Hot-rolled steel—Blue-black in color and scaly in appearance. It produces medium length, straight white sparks with only a few spurs. (C) Cold-rolled steel—Black or silver in color with a smooth surface finish. It produces medium length, straight white sparks with only a few spurs. (D) High-carbon steel—Black or silver in color. It produces short white to yellow sparks with lots of spurs. (E) High speed steel—Black or silver gray in color. It produces long dark, red sparks with spurs at the ends. (F) Stainless steel—Silver in color. It produces long light-colored sparks with a few spurs.

samples of known alloys. The key to good identifications is the use of consistent grinding pressure on the wheel.

As a rule of thumb, remember that most steels produce a large volume of sparks. Whether the steel is hardened or not makes little difference. The little *spurs* or spark explosions are caused by the carbon in the steel. The higher the carbon content, the more spurs you will see. Wrought iron, which contains very little carbon, has a long straight spark pattern with few spurs. High-carbon steel and cast iron, both of which contain a lot of carbon, produce short spark patterns full of spurs.

The color of the spark identifies other alloying elements. The sparks from plain carbon steel are almost white. Sparks from gray iron are orange, and sparks from high speed tool steel are dark red.

STAINLESS STEELS

There are three basic series of stainless steel—the 300 series (chromium-nickel series), the 400 series (straight chromium series), and the PH series (precipitation hardening series). Of the three series, the first two will be of most interest to the small project machinist.

300 Series Stainless Steel. These chrome-nickel types of stainless steel are probably the most common. They are used to make shafts, nuts, bolts, screws, food processing equipment, and marine equipment. They are used wherever corrosion resistance and toughness are primary concerns.

The 300 series stainless steels are considered nonhardenable and nonmagnetic. They do work-harden slightly, and in a work-hardened condition they are slightly magnetic. Work-hardened 300 series stainless can be annealed by heat treatment. These steels require heating to a bright red heat and rapid cooling or a water quench.

The 300 series stainless materials are available in sheets, plates, bars, and some structural shapes. The 303S and 303SU steels are classified as free-machining, but this term may be somewhat misleading. They are only free-machining when

compared to other 300 series materials. All stainless steels are tough and somewhat difficult to machine. The rating for 303SU is 60 percent compared to SAE 1112. All other materials in the 300 series are rated at 45 percent. Cutting speeds should be slow, and the use of cutting oil is recommended. The weldability of 300 series stainless is good, except for the free-machining alloys that are rated as only fair.

400 Series Stainless Steel. These stainless steels are heat-treatable by conventional heat treating methods. They are used where good mechanical strength is required and corrosive conditions are not too severe. These steels are used to make kitchen utensils, knives, valves, pump parts, aircraft parts, etc.

The 400 series stainless steels are magnetic. Like the 300 series, they are somewhat difficult to machine. Machinability is rated at 54 percent for most of the series, except 416 stainless. It is considered a free-machining material rated at 97 percent. All of the 400 series are easily welded, with the exception of 416. Welds in 416 are quite brittle and have a tendency to crack. The 400 series is readily available as sheet, plate, and bar stock.

NONFERROUS METALS

For commercial work, ferrous materials are specified for more applications than all other materials combined, but for small shop projects the majority of the work may be done with nonferrous materials. The relative ease with which aluminum and brass can be machined makes these materials favorites in the small shop.

Aluminum

Aluminum is lightweight, strong, easily machined, readily available, and comparatively inexpensive. All of these properties combine to make aluminum a very popular material. Pure aluminum is relatively soft and ductile. It has a tensile strength of about 13,000 psi. Aluminum is frequently alloyed with other elements to improve its

Table 9-2. Aluminum Alloy Groups.

Principal Alloying Element	Alloy No.	Comments
Pure aluminum	1XXX	Non-heat treatable
Copper	2XXX	Heat treatable
Manganese	3XXX	Non-heat treatable
Silicon	4XXX	Heat treatable
Magnesium	5XXX	Non-heat treatable
Magnesium and Silicon	6XXX	Heat treatable
Zinc	7XXX	Heat treatable
Other	8XXX	

mechanical properties. Like steel alloys, aluminum alloys have four digit numbers that identify them. The first digit identifies various alloy groups by identifying the principal alloying element. The following digits identify specific alloys within each group. Table 9-2 shows the basic alloy groups.

Pure aluminum and certain aluminum alloys are not hardenable by heat treating. These alloys do work-harden, and cold working can almost double the strength of some alloys. Aluminum temper conditions are specified by a letter and a one or two digit number following the alloy number. The letter designates whether it is in a work-hardened condition or a heat-treat condition. The number specifies the temperature. These designators are further explained in Table 9-3.

The machinability standard for aluminum alloys is 2011. This particular alloy is the best machining alloy of all the aluminum alloys. It is rated at 100 percent in both the T-3 and the T-8 conditions. This alloy is commonly used for fabricating complicated parts requiring extensive machining. It is weldable, has good corrosion resis-

tance, and good strength and hardness. It is readily available as bar stock.

One of the best known aluminum alloys is 2024. It has high strength and good fatigue resistance. This alloy is commonly used in structural applications where good strength-to-weight ratios are desired. Alloy 2024 can not be welded. Machinability in the T-4 condition is 90 percent of 2011; in the 0 condition it is rated at only 50 percent. The drop in machinability is due to the gumminess of the material in the softer condition. This is a typical characteristic for all aluminum alloys—the harder tempers in each alloy are easier to machine. The 2024 alloy is readily available in sheet, plate, bar, and rod.

Another well-known aluminum alloy is 6061. This is the least expensive and most versatile of the aluminum alloys. It has good corrosion resistance, can be welded, and has excellent mechanical properties. Machinability is rated at 75 percent of 2011 in both the T-4 and the T-6 conditions. Like 2024, it is available in sheet, plate, bar, and rod.

Heat treating processes for aluminum are somewhat complicated. For this reason, aluminum alloys are usually purchased in the desired heat-treat condition. If for some reason it is desirable to change the temper of a part after machining, the process is best left to a commercial heat treater. Welding does affect the temper of aluminum alloys. Unless a welded part is annealed and reheat-treated, the area adjacent to the weld is subject to cracking.

Table 9-3. Aluminum Alloy Temper Designations.

F	As fabricated	T-1	Heat treated to a specific temper. The "T" is followed by a number from 1 to 6 with the softest condition starting at 1 and graduating to the hardest condition at 6.
O	Annealed to softest condition		
H	Strain hardened, usually followed by one or more digits to indicate degree of hardness.		
H-1	Strain hardened only	T-7	These numbers indicate a T-6 condition with additional work hardening. T-10 is the hardest condition possible.
H-2	Strain hardened and partially annealed.		
H-3	Strain hardened and stabilized.	T-10	

Brass

Brass belongs to the family of copper alloys. It is made up primarily of copper and zinc and may also contain a small percentage of lead. It is probably the most popular material among machinists because it is easily machined, polishes well, and has a rich, attractive color. Most brass alloys can be soldered and brazed, but they do not weld. They are not hardenable by heat treating, but they do work-harden.

Like aluminum and steel, there are many different brass alloys, each formulated for certain specific purposes. The most popular of these alloys is probably alloy 360. This alloy is a free-machining brass used as the machinability standard for all copper alloys. Like 1112 steel and 2011 aluminum, it is rated at 100 percent. Typical uses for alloy 360 include various machined parts like gears.

Highly leaded brass alloy 3532 was once known as clock brass. It has a higher zinc content than alloy 360 and is considerably harder, especially in the work-hardened condition. Typical uses for this alloy include clock parts and bearings. Machinability is 90 percent of alloy 360. Both alloys 360 and 3532 are available in sheet, bar, and rod stock.

Muntz metal, alloy 3711, is similar to alloy 3532. It is a high strength material that has a machinability rating of 90 percent of alloy 360. This material is a favorite for hinges, locks, and a variety of cast and extruded parts.

Another very common brass alloy is *naval brass*, alloy 464. This brass is formulated specifically to resist saltwater corrosion. It is harder than alloy 360, but not quite as hard as alloy 3532. It is used to make brass hardware such as nuts and bolts, pump parts, and marine fittings. Its machinability rating is only 30 percent of alloy 360. Naval brass is available in bar stock only.

Bronze

Bronze is made primarily of copper and tin. These materials are generally harder and stronger than brass. Typical applications for bronze include propeller shafting, threaded fasteners, marine hardware, and valve parts.

Bronze is another metal that is not hardenable by heat treating, but is subject to work-hardening. These alloys can be soldered and brazed, but are difficult to weld. The machinability rating for bronze ranges from 30 percent for silicon bronze to 90 percent for architectural bronze.

PLASTICS

Plastic is a term that covers an extremely wide range of materials. Out of the many types of plastic available, there are only a few that are of interest to the machinist. These few include plexiglass, polycarbonate, nylon, and polytetrafluoroethylene.

Acrylic. The plastic that most people know as plexiglass is actually acrylic. (Plexiglas is just one manufacturer's registered trade name for acrylic material.) Acrylic plastic is available as sheet, plate, bar, and tube stock, as a colorless, transparent material, or in a number of translucent and semi-opaque colors. Typical uses for this material include display cases and decorative items.

Acrylic plastics can be machined by almost any conventional method, but two prime concerns are that it is a brittle material that fractures easily, and it is a thermoplastic, which means that it softens when it gets hot. These concerns are not major problems as long as certain precautions are taken.

To prevent acrylic sheet from cracking, extra care should be taken to make sure it is properly supported and secured during all machining operations. If sawing the material with a handsaw, clamp it between two boards (Fig. 9-16). The boards should support on both sides of the cut as close to the cut as possible. After cutting, the cut edges should be filed or machined smooth to eliminate stress areas from which cracks may start.

The fact that acrylic is a thermoplastic and softens when it gets hot means that you need to use relatively slow cutting speeds to prevent it from overheating. A stream of cooling air from a fan or an air compressor will help to keep it cool, or a small amount of cutting fluid can be used to help dissipate the heat.

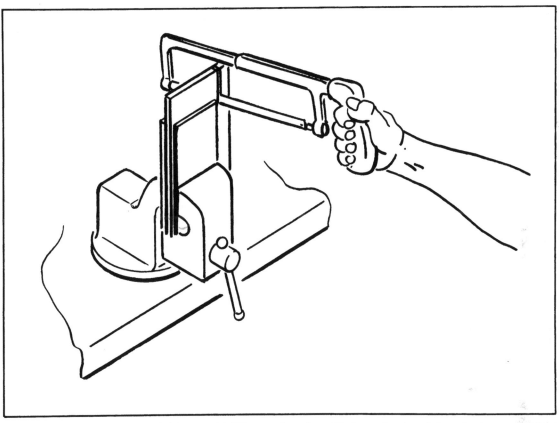

Fig. 9-16. Acrylic plastic is a very brittle material. When sawing, clamp it between two boards in a vise to prevent it from cracking.

Polycarbonate. Polycarbonate is another material that is best known by its registered trade name of Lexan. This material looks and behaves much like acrylic, except that it is a very tough material capable of withstanding a lot of harsh use. Like acrylic, it is available as sheet, plate, bar, and tube stock in many colors. Polycarbonate can be used in the same applications as acrylic. The cost is somewhat higher, but when you require a material that is less brittle or capable of withstanding higher temperatures, polycarbonate is the answer.

Nylon. Nylon is a valuable material to have around a machine shop. It is lightweight and easy to machine. Its toughness, low friction, and good abrasion resistance make it an excellent choice for mechanical parts that are subject to wear. Light bearings, sliding parts, and gears are all examples of the types of parts that can be made from nylon.

Nylon is available as plate, block, rod, and tube stock. Its natural color is white. Sometimes nylon may be mixed with fillers such as carbon-black or molybdenum disulfide to alter its properties. Carbon-black increases its tensile strength and hardness, and makes it less sensitive to aging in bright sunlight. Molybdenum disulfide increases its lubricity and reduces its coefficient of friction.

Polytetrafluoroethylene. Teflon is a registered trade name for polytetrafluoroethylene (TFE). This material is an excellent bearing material with a lower coefficient of friction and more lubricity than nylon. Somewhat more expensive than nylon, it is available in sheet, bar, rod, or tube.

195

Chapter 10

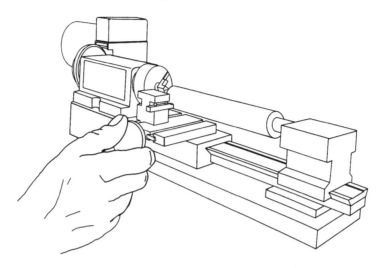

Heat Treating and Finishing

PARTS MADE FROM MATERIALS SUCH AS BRASS and plastic require no more than the removal of burrs and perhaps a little polishing before they are ready for use. These materials are not heat-treatable, so other than a light coat of wax or lacquer to protect their finish they require no special treatment. Steel parts, on the other hand, are heat-treatable, and they may also require the application of some kind of finish to prevent them from rusting.

Aluminum parts are somewhere between these two extremes. You generally purchase aluminum in the temper that you want the finished part to have, so there is no need for heat treating. To make aluminum parts look good and to give them a good measure of corrosion resistance, though, there are some commercial finishes you may want to consider.

An amateur machinist will need to know the various processes of finishing, those that can be done in the small workshop, and those that will have to be done commercially.

FINISH MACHINING AND POLISHING

After completing the machining of a part on a mill, lathe, or drill press, it is not unusual to find edges that have burrs or are unnecessarily sharp. No good machinist would ever call a part finished until he had removed these sharp edges. A countersink does a fine job of breaking the edge around a drilled hole. A fine file or a piece of fine emery cloth can be used to remove the rest of the sharp edges. How much you round or chamfer an edge depends on the part, but the corner on a finished part should not be sharp enough to cut you during normal handling.

After a part has been deburred, you may want to put a uniform surface texture over the entire part. The desired texture depends on how the part is to be used. If it goes inside an assembly and cannot be seen, you probably won't care too much about looks. You may want to polish small, decorative items. On larger items, you may just want to disguise small surface blemishes and scratches without polishing.

The easiest way to disguise small surface scratches or blemishes on parts with large flat surfaces is to use an orbital sander (Fig. 10-1). The idea here is not to remove material, but just to create a uniform surface texture that will make small scratches less obvious. The resulting surface roughness is dependent on the grit of the emery paper that is used in the sander. Usually a fine grit paper (180 grit) will produce a good looking finish.

Another method that can be used is a light

Fig. 10-1. A vibration sander can be used to apply a uniform surface finish that will hide scratches and other blemishes.

sandblasting. This is particularly good for small parts and parts that have curved surfaces. The cost of a small canister-type sandblast gun is not too great, especially if you already have an air compressor. Remember to always wear a good set of safety glasses for sandblasting. The actual work is best done outside and away from your machinery. Dust from the sand will cover everything if done inside, and the dust is very abrasive.

Polishing is accomplished by successively working the part with finer and finer abrasive materials. The coarsest abrasive used should be no coarser than required to quickly remove the worst surface blemish on the part. An orbital sander can be used initially. Finish polishing with a 600 grit emery should be done by hand. The polishing should be done with a uniform pressure and all strokes made in the same direction. A little oil used with the emery will help to produce a good polish. See Table 10-1.

If the part is cylindrical, the polishing can be done quite easily in a lathe, but you should cover the ways on the lathe to prevent any abrasive material from getting on them. Even if you keep the ways

Table 10-1. Abrasive Grit Size.

Extra Fine	Fine	Course
600	180	40
500	150	36
400	120	30
360	100	Extra Course
320	Medium	24
280	80	20
240	60	16
220	50	12

Note: Aluminum oxide is the preferred abrasive for hard metal polishing. Emery is best for softer metals. Crocus cloth, not shown in the table above is used for fine polishing and finishing on all metals.

covered, its a good idea to completely clean the lathe after finishing your work. The emery cloth can be torn in convenient width strips, held with two hands, and worked back and forth over the part.

Polishing may also be accomplished with a buffing wheel (Fig. 10-2). The abrasive or cutting compound is applied to the wheel by holding a stick of it against the wheel. When a wheel has been used with a particular compound, it should be marked and

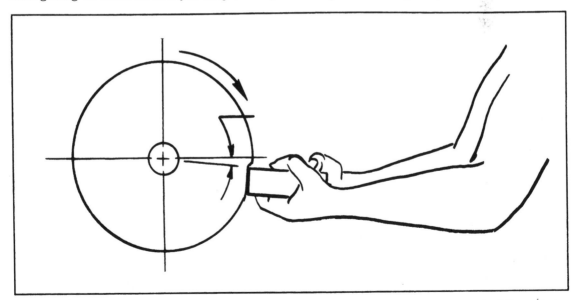

Fig. 10-2. When buffing, the part should be held firmly against the wheel slightly below center. The part should be worked up and down and sideways to blend the cuts.

used only with that compound in the future. Wear gloves whenever possible while buffing because the parts frequently become uncomfortably hot. Hold your work firmly against the buffing wheel, slightly below center, and work the part up, down, and sideways to blend the cuts. Inspect the work frequently and keep a tight grip on it. A fast moving wheel can snatch the work right out of your hands if you are not careful.

In any polishing operation, clean the part thoroughly when you change from one grit abrasive to the next. This will prevent a piece of the coarser grit from contaminating the finer grit and producing an unwanted scratch on the newly polished surface.

The polishing processes are all accomplished with sharp abrasives that actually cut and remove metal. After the polishing is completed, the final process called *coloring* is accomplished with a soft compound that brings out the natural color and luster of the metal. Coloring is always done with a buffing wheel. The same techniques used for polishing with a buffing wheel also apply here.

When selecting a buffing wheel, select the size from Table 10-2. The best size to use is dependent on the power and speed of the motor. Spiral-sewn wheels are stitched in a series of concentric circles from the center out to the face (Fig. 10-3). These wheels are hard and suited for cutting and polishing. Cushion-sewn wheels (Fig. 10-4) have only two or three rows of stitching to allow a more resilient cushioning effect. They can be used for either polishing or coloring. The softest wheels have only one row of stitching at the center. They generally are made from soft flannel. These wheels are in-

Fig. 10-3. Spiral sewn buffing wheels are hard and best suited for cutting and polishing.

Table 10-2. Buffing Wheel Selection.

Motor Size H.P.	Recomended Wheel Thickness			
	4″ Dia.	6″ Dia.	8″ Dia.	10″ Dia.
⅛	1″	½″		
¼	1½″	1″	½″	
⅓	2½″	2″	1½″	½″
½	3″	2½″	2″	1½″

Note: The best buffing speeds are from 5000 to 6000 surface feet per minute (SFPM). To calculate, divide the diameter of the wheel by 4 and multiply by the motor R.P.M.

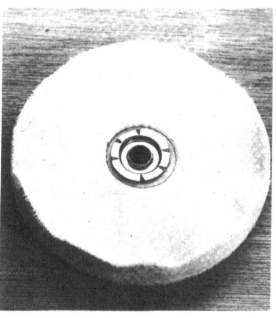
Fig. 10-4. Cushion wheels are soft and excellent for either final polishing or coloring.

tended for coloring only. Table 10-3 lists polishing and buffing compounds.

HEAT TREATING STEEL

Medium and high-carbon steels can be heat treated to either harden or soften them. The processes are simple and can be accomplished quite easily even in a small shop. Alloy steels and tool steels may require variations to the standard procedures, and the metal supplier's recommendations should be followed when heat treating them.

Annealing or softening steel is accomplished by heating the metal to its critical temperature and then cooling it slowly. Hardening steel is accomplished by heating to the critical temperature and cooling rapidly. At the critical temperature, approximately 1350° F., the molecular structure of the steel begins to change. This change continues as the temperature increases. Depending on the carbon content of the steel, the change is completed at a temperature somewhere between 1400° F. and 1700° F. The new molecular structure is called *austentite.*

When the steel is cooled, the molecular structure changes again. If the steel cools slowly, the new structure will be *pearlite.* This is the structure of annealed or softened steel. If the metal is cooled rapidly by quenching, the new molecular structure will be *martensite.* This is the structure of hardened steel. The actual hardness of the hardened steel will depend upon the carbon content.

High-carbon steels in the hardened condition are very hard and strong, but they are also quite brittle. A third heat treating process, *tempering,* reduces the brittleness. In tempering the steel is heated just enough to allow some of the molecular

Table 10-3. Polishing and Buffing Compounds.

Compound	Color	Uses
Tripoli	Brown	Fast cutting compound for removing scratches and pits on most metals.
Crocus	Dark Red	Cutting compound suitable for use on all metals.
Rough	Red	Cutting and polishing compound for soft metals such as gold, silver, and brass.
Rough	White	Coloring compound suitable for all metals.

Table 10-4. Critical Temperatures for Heat Treating Carbon Steel.

Carbon Content	Temperature In Degrees F.	Incandescent Color
0 - 0.12%	1650° ± 50°	Cherry Red
0.12 - 0.30%	1575° ± 25°	Cherry Red
0.30 - 0.50%	1525° ± 25°	Dull Cherry Red
0.50 - 1.00%	1475° ± 25°	Dull Cherry Red

structure to change to pearlite. This process is perhaps the most important part of the heat treating sequence. It permits you to tailor the characteristics of the steel to meet your needs. A hammer made from high-carbon steel and fully hardened would be too brittle to use. Tempered to reduce its brittleness, it takes on some of the toughness of annealed steel while still retaining most of its hardness.

Professional heat treaters use some very expensive and sophisticated equipment to control their processes. Precise temperatures and special quenching solutions are used depending on the type of material, the size of the part, and the desired result. With unsophisticated equipment the results of heat treating will not be as accurate, but using the methods passed down by blacksmiths for many years, you can achieve surprisingly good results.

Annealing

As previously stated, the steel is heated and cooled slowly during annealing. The upper limit of the critical temperature is dependent primarily on the carbon content of the steel. The less carbon the steel contains, the higher the upper limit of the critical temperature. See Table 10-4. Using the methods developed by blacksmiths, note the color of the metal as it progresses through its incandescent stages. High-carbon steels have been heated sufficiently when they reach a dull, cherry red glow when viewed in normal indoor light. Low-carbon steels need to be heated to a full cherry red. Medium-carbon steels reach the upper limit of their critical temperature somewhere in-between.

Hold the part at temperature long enough to ensure that the heat has been distributed uniformly throughout the part. Parts with large cross-sectional thicknesses will require more time than thinner parts. After the part is thoroughly heated, it is cooled very slowly. If the heating was done in an oven, just turn the oven off and allow the part to remain in the oven until cool. If the part was heated with a torch, remove the flame gradually. The initial cooling period is the most critical. After the

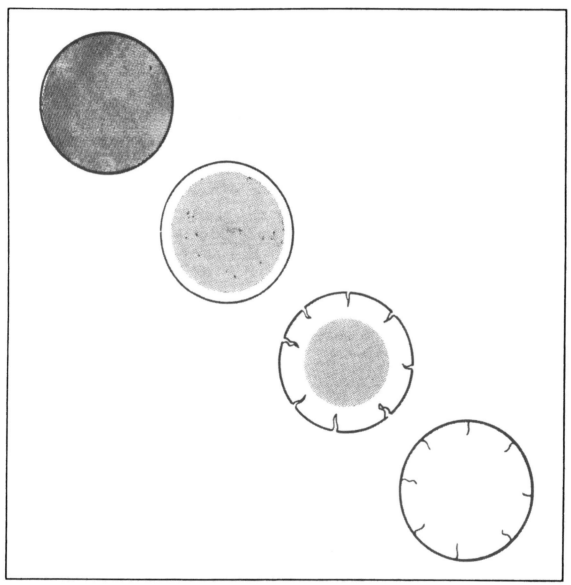

Fig. 10-5. All materials expand when heated and contract when cooled. If a large part is cooled too rapidly, as illustrated above, the surface material will shrink around the still expanded center material. Unable to stretch, small surface fractures may develop. As the core of the material cools and shrinks, the cracks will close and become almost invisible to the naked eye, but they will still be there and they will weaken the part. For this reason large parts should be quenched in warm or hot oils to slow down the quenching action.

temperature has dropped to 900° F. or less, the cooling time is no longer critical and the part may be quenched.

Hardening

The only difference between the annealing process and the hardening process is the way the part is cooled. In either case you need to heat the part to the upper limit of its critical temperature. After heating as thoroughly as possible, the part is cooled rapidly to a temperature below 400° F.

Cooling can be accomplished by quenching in water or oil. Water is used for most small parts. For large parts the water quench may be too severe because as the metal cools, it shrinks. If the part is large, the outside will cool much more rapidly than the inside. This means that the outside will shrink faster than the inside. Unable to compress the inner material, the surface will start to split, creating small surface cracks (Fig. 10-5). To prevent this, the part may be quenched in a hot oil that will not conduct the heat away quite as rapidly as water would.

When quenching, the part should be submerged as quickly as possible. If the part is a rod or plate, it should be immersed vertically into the quenching bath because if one side cools more rapidly than the other, it will shrink and cause the part to warp (Fig. 10-6). By holding the part vertically as you immerse it, you make sure that both sides cool at the same rate with less chance of warpage. When the part has been hardened, it will be extremely brittle and impossible to straighten unless annealed again.

As the part is immersed into the quenching bath, it should be swirled around to prevent steam or gas pockets from forming and causing an uneven rate of cooling. This is especially important for irregularly shaped parts that might trap steam bubbles because of their irregular shapes.

Case Hardening

The previously stated process for hardening steel works for medium and high-carbon steels only. Low-carbon steels do not contain sufficient carbon for hardening, but *case hardening* is a process that can be used to create a hard surface on a low-carbon steel part. To accomplish this process, it is necessary to convert the surface of the part to high-carbon steel by packing the part in a carbonaceous material and then heating it to approximately 1700° F. (a bright cherry red glow). When the part gets hot, the carbon will penetrate into its surface. The actual depth of penetration may vary

Fig. 10-6. Immersing a hot part into the quenching bath at an angle will cause one side of the part to cool faster than the other. The uneven cooling may cause the part to warp. Always submerge the part vertically into the quenching bath.

203

from 0.001 to 0.050 inch depending on how long the heat was maintained.

After the part has been held at temperature, it is quenched in the same manner that a part made from high-carbon steel would be quenched. The outer surface of the part will be just as hard as if it were made from high-carbon steel. The core of the part will retain the properties of low-carbon steel. The advantage of this process is that the finished part will have the toughness, flexibility, and resistance to impact and fatigue that a low-carbon steel would have. The part will also possess the hardness and wear resistance of a hardened high-carbon steel part. Parts that are generally case-hardened are gears, crankshafts, and some gun parts.

There are a number of commercially prepared case hardening compounds available, or you can make your own from powdered charcoal. The compound should always be packed tightly around the part to prevent any voids or air spaces; it has to be in physical contact with the part in order to work. The part should be maintained at temperature anywhere from ½ hour to 2 hours to get good penetration. Pack some scrap pieces of low-carbon steel in with your part to be used as test samples. You can pe-riodically remove one, quench it, and then test it with a file for hardness and depth of penetration.

Tempering

Tempering is the process that permits you to tailor the hardness of a part. This is the most critical of the heat treating processes. Slight variations in temperature can make large differences in the resulting characteristics of the steel. Fortunately, nature provides a good indicator of temperatures within the tempering range. The surface color of steel changes rapidly as the temperature increases, and the color changes occur at the critical tempering temperatres. Table 10-5 shows tempering temperatures, the corresponding colors, and the recommended tempers for various parts.

Tempering should take place as soon after hardening as possible. The first step in tempering a part is to clean and polish one or more surfaces so that they are bright and shiny. After polishing, slowly heat the part until it takes on the desired color, then quench it. The reason for heating the part slowly is to minimize the risk of overheating and tempering it too much. Also, slow heating provides a more uniform heating and permits more

Table 10-5. Typical Tempering Temperatures and Colors.

°F.	Temper Color	Typical Parts
380	Very Light Yellow	Lathe tools, lathe centers, scrapers, burnishing tools, scribes.
425	Light Straw	Metal working hammers, milling cutters, drills reamers
465	Dark Straw	Punches, broaches, hacksaw blades, taps, wood gravers, knife blades
490	Yellow Brown	Carpentry hammers, wood chisels, center punches, prick punches
525	Purple	Wood augars, stone cutters, drifts, awls, needles
545	Violet	Axes, cold chisels, skinning knives, molding cutters, springs
590	Pale Blue	Dental tools, screwdrivers, wrenches, wood saws

time for the molecular changes to take place. Many professional heat treaters repeat the tempering process twice. They claim that this results in tougher and better parts.

HEAT TREATING IN THE SMALL SHOP

Now that you have learned about the four basic heat treating processes, you should consider actually doing some work in your own shop. The equipment that you will need includes a means of heating the part, a container for holding the quenching medium, and some emery for cleaning the part. For small parts you can use a Mapp gas torch or a blowtorch. Propane torches do not get hot enough. I have even heated small parts on the kitchen stove. Larger parts are difficult to heat unless you have a furnace.

Actually purchasing a furnace would be expensive, but you can make your own furnace quite easily. To heat your furnace, you will need a forced air natural gas burner. You can make a burner with a few short pieces of iron pipe and an old vacuum cleaner. Plans for a forced air burner that will provide temperatures up to 2200° F. are shown in Fig. 10-7.

The furnace itself can be made by simply stacking firebrick to form a chamber (Fig. 10-8). To keep the cost down, they can be stacked directly on the ground or, better still, on a small metal table. Don't cement the bricks; just stack them so that the shape of the furnace can be rearranged to accommodate a variety of different sized and shaped parts.

Don't forget to build a firebrick floor in your furnace, especially if you build it on a table. The bricks won't prevent the table top from getting hot, but they will help to protect it. Never build your furnace on a concrete slab because moisture in the cement may cause it to crack and ruin the slab. (An electric oven is shown in Fig. 10-9).

For your first attempt at heat treating, let's assume that you have just machined a new center punch and now you want to harden it. The punch is made from a high-carbon steel, so case hardening isn't necessary. Before you actually start the heat treating process, let's consider which parts of the punch need to be hardened. The point has to be hard enough to punch hard metals without losing its sharpness. It can't be too hard or it will be brittle. Table 10-5 indicates that punches are usually tempered at 490° F., which is indicated by a yellow-brown color. Neither the body nor the head of the punch need to be hardened. (The head should not be hardened because a blow from a hammer might cause it to chip.)

The process will permit you to both harden and temper the punch with just one heating. The punch is small enough to be held with a pair of pliers and heated with a Mapp gas torch to a dull red glow (1475° F., for high-carbon steel) from the point back about 2 inches. As soon as the punch reaches temperature, it should be plunged to approximately half of its length into a container full of warm water. Swirl the punch around in the water to prevent steam bubbles from forming. When the tip has cooled sufficiently to stop boiling the water, pull it out and clean the pointed end with a piece of emery cloth. You only have to clean a narrow strip of metal to observe the color changes. You will have to work fast. The end opposite the point will still be hot, and the heat will be traveling down the length of the punch toward the point. As the heat travels into the area you have just cleaned, it will oxidize the surface and change its color.

You will see several distinct bands of color traveling toward the point. The first color will be a straw color, followed by bronze, purple, and finally light blue. Quench the punch by completely submerging it just as the yellow-brown color reaches the point. After quenching, examine the point. It should still have the same color. If you did not act fast enough and the point got too hot, it will be evidenced by one of the other colors. If the point did get too hot, it will be too soft. The entire process will have to be repeated.

If the heat retained in the body of the punch was not sufficient to heat the point to the proper color, you can play your torch over the body to give it more heat. Don't attempt to heat the pointed end

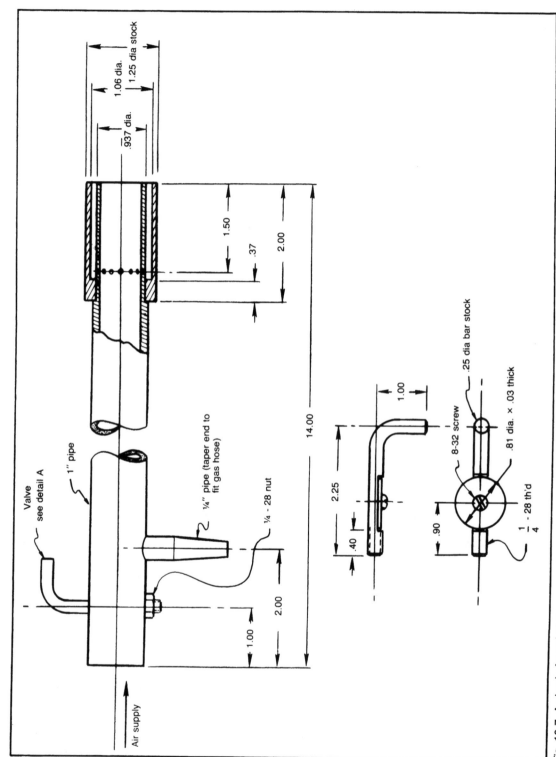

Fig. 10-7. A simple forced air burner can be made from iron pipe.

Valve
see detail A

1" pipe

¼" pipe (taper end to fit gas hose)

¼ - 28 nut

Air supply

1.00

2.00

14.00

.937 dia.

1.06 dia.

1.25 dia stock

1.50

.37

2.00

1.00

2.25

.40

.90

8-32 screw

.25 dia bar stock

.81 dia. × .03 thick

$\frac{1}{4}$ - 28 th'd

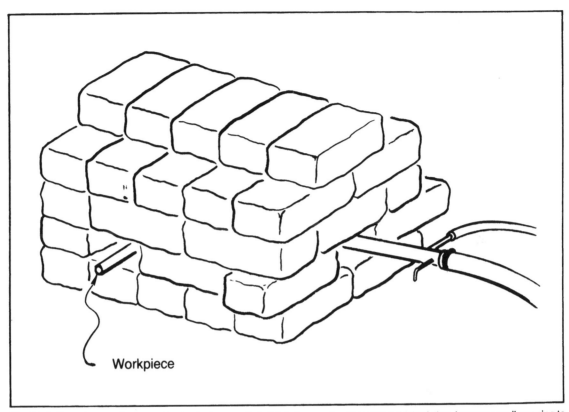

Fig. 10-8. To make your furnace, simply stack some firebrick to form a chamber of the desired size. Leave a small opening to insert the burner and another for inserting the part to be heated.

directly, or you will surely overheat it and have to start over again. If you have done the process properly, your punch will have a hard durable point that should last a long time. The rest of the punch will be in the annealed state so that it will have plenty of toughness to resist chipping during use.

If you want a part to have the same hardness throughout, quench it completely on the first quench. Clean the entire part so that you can observe the color changes. Then reheat it very slowly until it turns the proper color.

When the heat treating has been completed, the part can be given a final cleaning and polishing, then it will be ready for use. If you intend to polish the part, you should do most of the polishing prior to the heat treating. Hardened steel will be a lot harder to polish than annealed steel.

FINISHING AND COLORING METALS

There are several good finishes that can be used on steel parts to prevent rust and provide protection. The most widely used finishes are various types of plating. Usually the part is plated with an inert metal such as cadmium, nickel, or chrome. The plating processes can be accomplished in the small shop, but I believe the need for plating is too small to justify the expense of the required chemicals and equipment. For this reason, I recommend that if you have parts that you want to plate, you should have them done commercially.

Aside from plating, there are several attractive finishes that can be applied easily using methods that have been passed down by blacksmiths, gunsmiths, locksmiths, and other craftsmen for many years. No matter which method of finishing

207

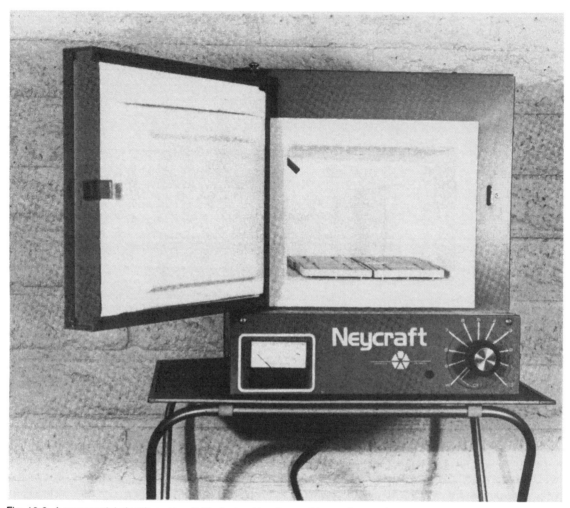

Fig. 10-9. A commercial electric oven suitable for heat treating and tempering steel parts.

you use, it won't hide surface defects. If you want the parts to look good, polish them before you apply the finish.

Many of the finishes described in the remainder of this chapter are chemical finishes. You will be working with materials that can be dangerous. Many of the chemicals are poisonous, some are highly corrosive, and all require special care and attention.

All chemicals should be stored in dark glass containers, clearly labeled and kept out of the reach of children. Always work in a well-ventilated area.

Wear eye protection and rubber gloves whenever working with acidic or caustic materials. Caution: it is one of the basic rules in mixing chemicals to always add acid to water. *Never mix water into acid.* Most of the chemicals used in the following formulas can be purchased through your local druggist.

Prior to applying any kind of finish to a part, the part must be thoroughly cleaned. The parts should be degreased in a good solvent, such as trichloroethylene, or by washing them in a solution of trisodium phosphate (TSP). You can purchase TSP at your local hardware store. Wear rubber gloves during cleaning operations. Don't handle the part with your bare hands after they have been cleaned. After degreasing, the parts should be rinsed in clear water and dried.

Heat Bluing Steel. Parts that are not heat-treated may be colored by heat bluing. Simply heat the part slowly and uniformly. The heating can be done by almost any method, a torch, furnace, and even an electric hot plate can be used for small parts. The desired temperature is 650° F. A slightly lower temperature will result in a lighter blue; a slightly higher temperature will result in a darker blue. After heating, the part should be coated with oil to prevent rusting.

Cold Bluing Steel Parts. If heating the part is impractical because of size or heat-treat condition, you can make a cold bluing solution that does a fine job of coloring steel. To make the solution, dissolve the following ingredients in 5 ounces of water:

> *2 ounces ferric chloride*
> *2 ounces antimony chloride (poison)*
> *1 ounce gallic acid*

The solution should be applied uniformly with a pad of glass wool and then exposed to the air for 24 hours. Then the part should be washed in hot water and air dried. To protect the part from rusting, it should be coated lightly with oil.

Parts may also be blued by using various commercial gun bluing solutions. These solutions may be purchased at almost any sporting goods store.

Cold Browning Steel. This is an old formula used by gunsmiths to color gun barrels. To make this solution you will need:

> *3/4 ounce copper sulfate*
> *1 ounce mercuric chloride (poison)*
> *1/2 ounce nitric acid, concentrated (highly corrosive)*
> *1 ounce denatured alcohol*
> *1 ounce tincture of ferric chloride*
> *1 tincture of ethyl nitrate*

Dissolve the copper sulfate and mercuric chloride in 24 ounces of water, then mix in the remaining ingredients. Remember to add the acid into the solution rather than the solution into the acid. The solution is applied to the part using a pad of glass wool, after which the part is left to air dry for 24 hours. Following this, the part should be washed in clean water and dried. A light coat of oil will prevent rusting.

Commercially prepared browning solutions are also available. Like cold bluing solutions, they may be purchased from most sporting goods stores.

Black Finish for Steel Parts. The following method can be used to provide a jet black, rust-inhibiting finish on any steel part. This finish looks especially nice on nuts, bolts, and screws used in small mechanical assemblies. To obtain the finish, simply coat the part with motor oil and then place it in an oven or on a hot plate. Heat it to approximately 350° F. The oil will bake into the part in about 10 minutes leaving a hard, durable black finish.

Finishing Aluminum. Many aluminum parts are used without any surface treatment because most aluminum alloys are fairly resistant to corrosion. If the part will be exposed to a corrosive environment or if it will be used in an application where looks are important, you may want to apply a finish.

Two common finishes may be applied to aluminum. The first is called *chemical film treatment* (chem-film). The second is called *anodize*. Both procedures require the use of numerous chemical baths, and the equipment costs for the small shop are prohibitive. As with electroplating processes, these procedures are best left to the commercial processors.

Chem-film treating produces a yellow finish on aluminum. The primary purpose of chem-film is to provide a stable, oxidation free surface for painting, but because it stablizes the surface it has another benefit for parts that are not painted and are handled frequently. There is a light aluminum oxide that forms on most alloys. This oxide rubs off easily and leaves a dirty gray residue on your hands or whatever else with which it might come in contact. Chem-film treating stablizes the surface and prevents this condition.

Anodizing is the preferred finish for aluminum. It is more expensive than chem-film, but it provides an attractive, hard, durable finish that can be dyed almost any color for decorative purposes. Left natural, it will give the aluminum a light gray color. The thickness may be specified and is usually between 0.0001 and 0.0010 inch. Anodizing will help keep aluminum parts from corroding.

Imitation Anodize. You can create an imitation anodize finish by dipping the part in a solution of 1 or 2 tablespoons of lye dissolved in one pint of water. Caution: lye is a very corrosive chemical. After the part turns a whitish color, it should be rinsed in clear water. Parts treated this way can be dyed any desirable color by immersing in a solution of household dye.

Coloring Brass. Most parts made from brass look fine just the way they are. It is sometimes desirable to alter the color in order to antique the part or give it a richer look. The following solutions are used for these purposes.

Boiling parts made of brass in this solution will give the part a rich golden color. The solution is made up of the following chemicals:

2 ounces potassium nitrate (saltpeter)
1 ounce common table salt
1 ounce alum
1 ounce concentrated hydrochloric acid
24 ounces water

The solution is made by dissolving the dry chemicals in the water and then adding the acid. Boil the part in the solution until it reaches the desired color, then rinse and dry. To protect the part from oxidizing, coat it with a clear lacquer. I like to just dip the part in a solution of 1 part lacquer and 1 part lacquer thinner. The resulting coating is thin enough to not be noticeable and still sufficient to give the part adequate protection.

This solution will give brass a green color similar to that seen on many old statues. It is an excellent antiquing finish.

3 ounces iron chloride
1 pound ammonium chloride
8 ounces copper acetate (verdigris)
10 ounces common table salt
4 ounces potassium bitartrate
1 gallon water

This solution may be applied with a brush, or the entire part may be immersed in it. The length of time that the part is in contact with the solution will determine the intensity of the resulting color. When the desired color has been obtained, the part should be rinsed in fresh water and dried.

To darken brass or actually turn it black, immerse it in a solution of 1-ounce copper nitrate dissolved in 6 ounces of water. Keep it in the solution until it turns a nice copper color. Then dry it and heat it until it turns the desired color.

A second method of darkening brass is to just expose it to the fumes from household ammonia.

Passivation of Stainless Steel. Stainless steels do not require any special coating to protect them, but unless stainless steel is *passivated* after machining, it is subject to corrosion and rust. The passivation process simply removes all of the ferrous material from the surface, leaving only the chrome and nickel exposed. To passivate a part, immerse it in a 50 percent solution of nitric acid for 1 or 2 hours, then rinse with water and dry it.

210

Chapter 11

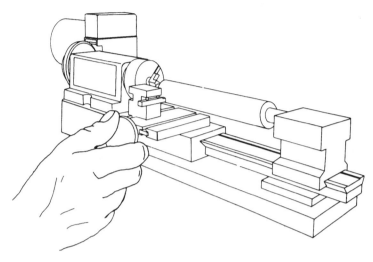

Projects

THE PROJECTS (FIG. 11-1) GIVEN IN THIS CHAPter have been selected for two reasons. First, they will give you some good experience in setup and machining techniques. All of the projects are fairly simple, but in completing them you will perform most of the operations that are called for in machine work. The confidence gained, not only in yourself as a machinist but also in your equipment, will prove valuable to you as time goes by.

Second, these projects have been selected because they are all fun projects. The tools are all items that will come in handy in your shop, and you will get a lot of satisfaction out of knowing that you made them yourself. I have never known a machinist who didn't have a number of homemade items in his tool chest. With the experience you gain in making these tools, you can improvise and make special tools to fill any need that might arise.

These projects are designed to be completed in sequence. With the exception of the last two projects, they all increase in complexity and in the degree of skill required to complete them. Operations that are explained in detail for one project are

not explained again for subsequent projects. For this reason, if you decide to skip projects, it might be a good idea to at least read through each project to become familiar with all of the operations.

As you make the projects, keep in mind that the dimensions are there to guide you. They are not critical unless specifically noted. You can change them and modify the part to suit any particular need you might have, but, if possible, hold the dimensions as accurately as you can. This will give you valuable experience so that when you do need to stay within certain tolerances, you will know how to do it.

PROJECT 1: CENTER PUNCH

Making your own center punch or prick punch (Figs. 11-2 and 11-3) is a fairly simple project that will help you to develop basic skills in lathe work, finishing, and heat treating. To get started you will need a piece of ¼-inch hex bar stock 4.06 inches long. You can use any high-carbon steel for this project. I like to make tools out of 4130 steel; it machines well, takes a good polish, and is fairly

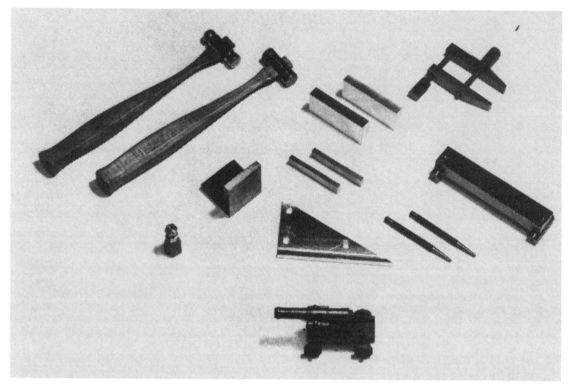

Fig. 11-1. Projects.

corrosion-resistant. If the material is in a hardened condition, you will have to anneal it before you start to machine it. The step-by-step procedure for making your punch is as follows:

1. Cut the bar stock 4.06 inches long. The extra 0.06 inch of length will give you ample material on both ends to take some cleanup cuts.

2. Using a three-jaw chuck, mount the material in your lathe with approximately 1½ inches of material sticking out of the chuck. Using minimal depth cuts, face the end of the bar.

3. Using a tool with a 0.03-inch nose radius, turn the end of the bar stock down to the 0.230 × 0.37-inch long diameter. This can be done in several cuts and the final cut should be a light one with a slow, uniform feed to achieve a good finish.

4. The 0.020-inch × 45° chamfer can be cut with a fine flat file. Before removing the part from the lathe, take a piece of fine (600 grit) emery cloth and polish the surfaces that you have just cut. A piece of paper over the bed of the lathe will prevent any abrasive material from getting on the lathe's slides.

5. Remove the part and rechuck it with the finished end in the chuck. Chuck it with approximately 1½ inches of material protruding out of the chuck. If you are using a Sherline lathe, rotate the headstock to 5° as shown. If you are using a lathe with a compound slide, rotate it to obtain a 5° taper. Then, using a series of cuts, turn the tapered end of the punch to obtain the 5° × 0.960-inch dimensions.

6. The 90° point is machined by rotating the headstock to 45° (Fig. 11-4). If you want to make the prick punch, rotate the headstock to 30° (Fig. 11-6). When these cuts are completed, use some emery cloth and polish the newly cut areas.

7. At this point the punch could be heat-treated, but it will look a lot nicer if the hex surfaces

212

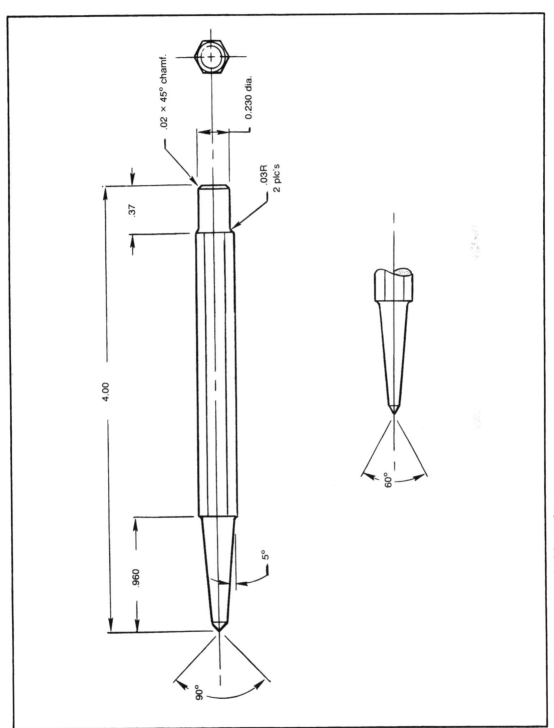

.02 × 45° chamf.

0.230 dia.

.03R
2 plc's

.37

4.00

.960

5°

90°

60°

Fig. 11-2. Project 1—center and prick punches.

213

Fig. 11-3. Finished punches.

are filed flat and polished. This is accomplished by clamping the part in a vise equipped with brass or aluminum jaw liners to prevent damaging the part. Each of the six sides of the hex should be draw-filed to remove any blemishes. Then they can be polished with a piece of fine emery cloth.

8. Heat treating can be done with a torch as described in Chapter 10. Don't forget to temper the point at 465° F (dark straw color). After tempering, the punch can be given a final polishing, and then it's ready for use.

PROJECT 2: MACHINIST'S JACK

Machinist's jacks are frequently used to support irregularly shaped parts during machining operations. The size can be varied, but the one shown in Figs. 11-8 and 11-9 makes a good universal tool.

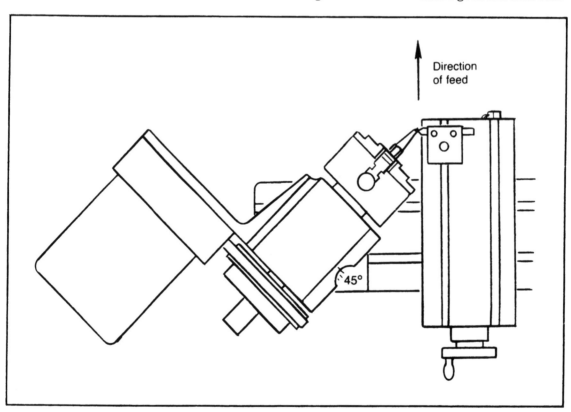

Direction
of feed

45°

Fig. 11-4. To machine the 90° center punch point on a Sherline lathe, rotate the headstock to 45° as shown. The arrow indicates the direction of feed.

Fig. 11-5. Machining the center punch point.

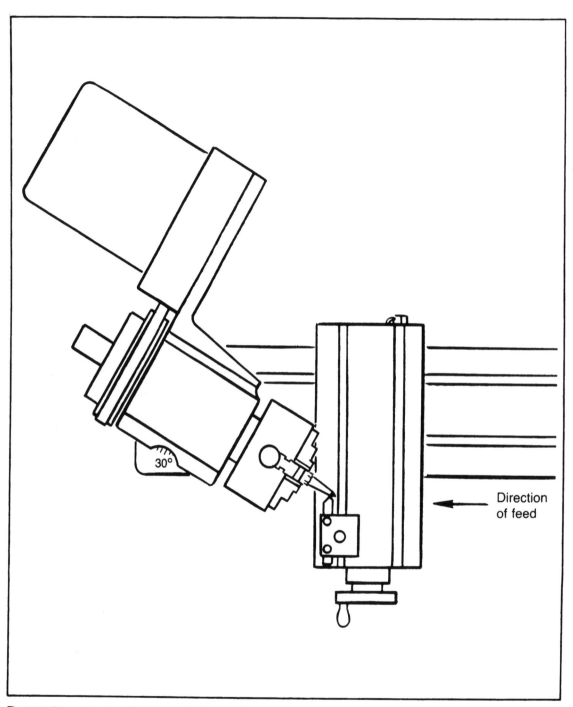

Fig. 11-6. To machine the 60° prick punch point on a Sherline lathe, rotate the headstock to 30° as shown. The arrow indicates the direction of feed.

Fig. 11-7. Machining the prick punch point.

To make yours, you will need a 5/16-24 × 0.75-inch long hex head bolt, two 5/16-24 jamb nuts, and a piece of .75-inch hex bar stock .87 inches long. Any kind of steel will do.

1. All of the operations required for machining the base are straightforward. Both ends of the stock need to be faced.

2. The stock is then drilled through with an I

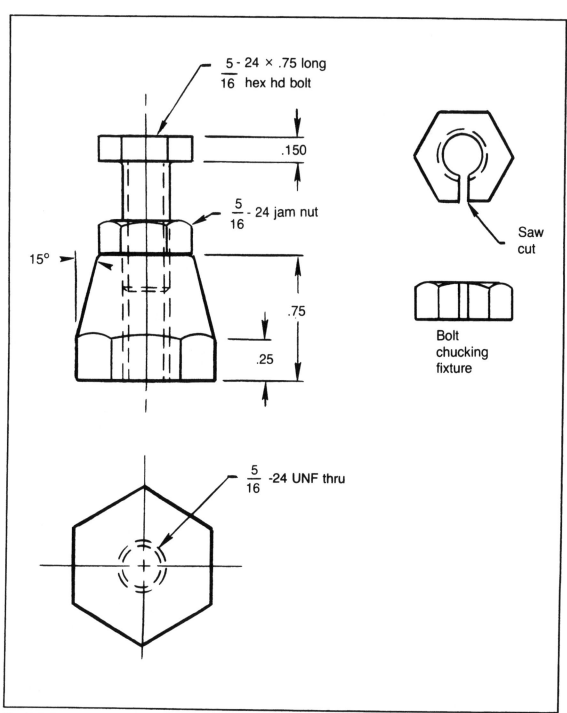

$\frac{5}{16}$ - 24 × .75 long hex hd bolt

.150

$\frac{5}{16}$ - 24 jam nut

15°

.75

.25

Saw cut

Bolt chucking fixture

$\frac{5}{16}$ -24 UNF thru

Fig. 11-8. Project 2—machinist's jack.

Fig. 11-9. Finished machinist's jack.

drill (.272-inch diameter) and threaded 5/16-24 UNF (Fig. 11-10). If the drilling is accomplished in a small lathe such as the Sherline, drill a pilot hole, ⅛- or 3/16-inch diameter, first. Then enlarge the hole to its full diameter.

3. Following the drilling and threading operations (Fig. 11-11), the 15° taper is machined. The part is then polished and set aside, pending completion of the remaining parts.

4. To make the jack screw, you simply need to machine the head of the 5/16-24 bolt flat. This is nothing more than a facing operation done in your lathe. The trick comes in chucking the bolt without damaging the screw threads. One of the simplest ways to do this is to saw through one side of a standard 5/16-24 nut. Now the screw can be chucked up in the nut, and the nut will grip the screw threads without damaging them.

5. After machining, the parts may be assembled. The jack is ready for use. A light coat of oil will help to prevent rust when the jack is not being used.

PROJECT 3: BRASS MALLET

The brass mallet (Figs. 11-12 and 11-13) is a simple project that will make use of both your lathe and your milling machine. To make the mallet, you will need a piece of brass stock 0.75-inch diameter × 2.00 inches long. The step-by-step procedure is as follows:

1. Chuck the stock in a three-jaw chuck and face it. Then, using a series of cuts, reduce the diameter of the stock to 0.700 inches for a distance of 1.50 inches.

2. Using a 60° threading tool cut a 0.062-inch deep V groove around the part 0.300 inches back from the free end of the stock. The 0.300-inch dimension can be established by aligning the tip of the cutting tool with the end of the part, then using the feedscrew micrometer dial to move the tool over the 0.300-inch dimension. The depth of the groove is established in a similar manner. Bring the tool in until it just touches the part, then use the micrometer dial on the crossfeed feedscrew to bring the tool in an additional 0.062 inch.

3. Use a fine flat file to cut a slight crown on the face of the hammer and also to cut the 0.020-inch × 45° chamfer around the face. Following this, polish all of the surfaces with a fine emery. Don't forget to cover the slides on the lathe when polishing.

4. After polishing, the part is removed and rechucked so that the opposite end can be turned. When reducing the diameter to 0.700 inches, be careful to blend the diameter with the previously cut diameter. Then repeat steps 1 through 3.

The next step is to cut the hole for the handle. If you don't own a mill or a milling attachment for your lathe, you can make the hole by drilling a series of holes and then filing them to produce the

Fig. 11-10. Drilling the machinist's jack prior to tapping.

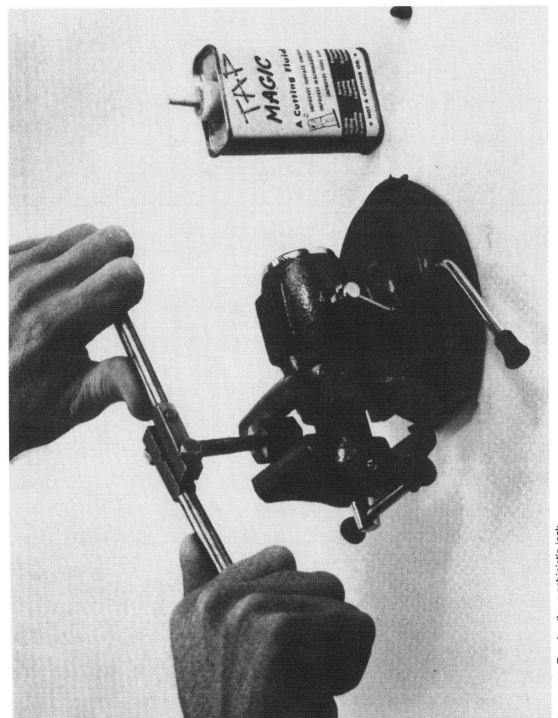

Fig. 11-11. Tapping the machinist's jack.

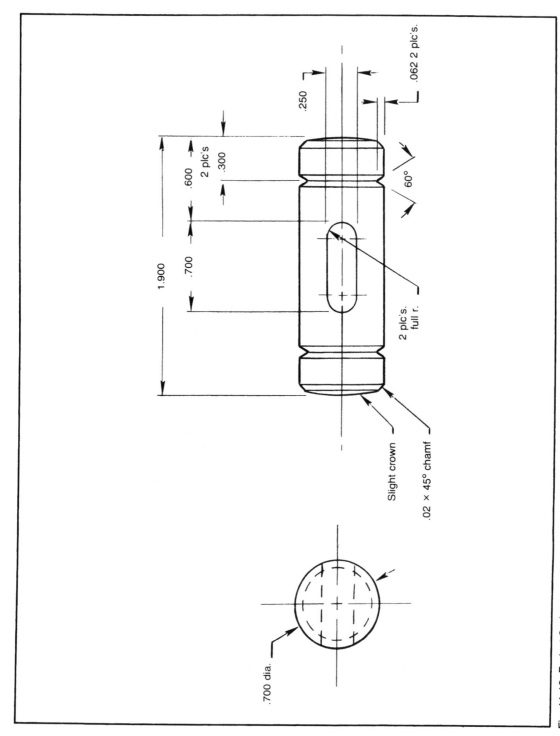

.062 2 plc's.

.250

60°

.300 2 plc's

.600

.700

1.900

2 plc's. full r.

Slight crown

.02 × 45° chamf

.700 dia.

Fig. 11-12. Project 3—brass mallet.

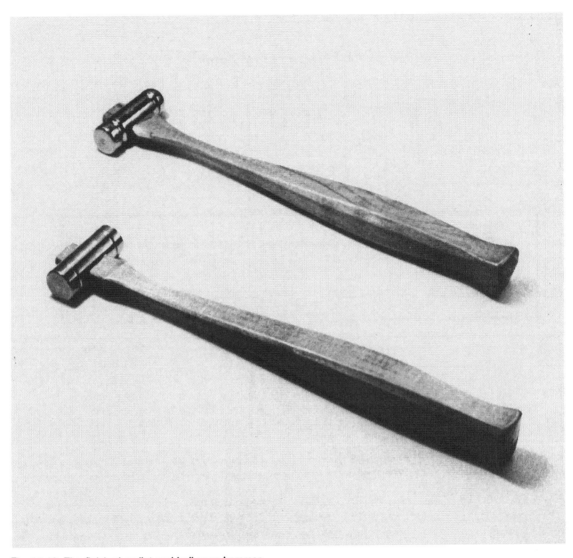

Fig. 11-13. The finished mallet and ball peen hammer.

slot. That would be doing things the hard way, however. You can do a much better job with a milling machine. Use a 0.250-inch diameter end mill ground for plunge cutting.

5. To set your mill up, mount your vise in the center of the table with the jaws parallel to the X axis (Fig. 11-14). When clamping the part in the vise, use a piece of scrap material to space the part away from the bottom of the vise so that there will be no chance of cutting into the vise during the slotting operation (Fig. 11-15). Clamp the part so that the face of the mallet protrudes past the edge of the vise by at least ¼ inch.

6. To locate the cutter in relation to the part, bring it over slowly until it just touches the side of the part. Turn the cutter by hand to make sure it is the true diameter of the cutter contacting the part. A piece of paper or thin shim stock held between the

223

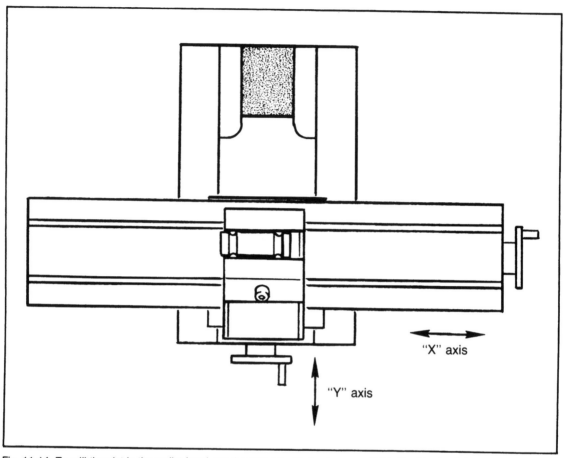

Fig. 11-14. To mill the slot in the mallet head, set the vise up with the jaws parallel to the X axis of the machine.

cutter and the part will prevent accidentally cutting into the part (Fig. 11-16). *Caution:* Do not attempt to do this with the machine running. Once the edge of the part has been found, the center position can be found by adding the cutter diameter to the part diameter and dividing by two. Center the cutter over the part using the micrometer dial on the Y axis feedscrew. From this point on, until the part is completed, there should be no more movement on the Y axis of the machine. To prevent any unwanted movement, the Y slide should be locked.

7. The next step is to locate the cutter in relation to the end of the part. Since the slot is dimensioned from the face of the mallet, start by locating the cutter against it. This is done by fol-

lowing a procedure similar to the one outlined in step 6 (Fig. 11-17). When the face of the mallet is located, the end of the slot can be located by adding the diameter of the cutter to the locating dimension. This location is found by using the micrometer dial on the X axis feedscrew.

8. Start your cut by plunging the cutter to a depth of approximately 0.050-inch and then moving the part on the X axis as calculated by subtracting the cutter diameter from the desired length of the slot.

9. At the end of the cut, back the feedscrew slightly to take up the play. Plunge the cutter another 0.050 inch deep and cut back in the opposite direction. This process is repeated until the slot is

cut completely through the part (Fig. 11-18).

10. When the slot is finished, the part can be removed from the mill and deburred. A handle can be purchased at most hardware stores or you can make your own, as I did, using a piece of scrap hardwood.

PROJECT 4: BALL PEEN HAMMER

The ball peen hammer project (Figs. 11-19 and 11-20) is similar to the mallet project except that it is designed to demonstrate the technique of lathe filing. You will need a piece of 0.75-inch diameter

× 2.06 inches long stock to make your hammer. The material must be a hardenable steel. Alloy 4130 chrome-moly steel will do fine. The step-by-step procedure is:

1. Chuck the stock in a three-jaw chuck and face it. Then, using a series of cuts, reduce the diameter of the stock to 0.625 inches for a distance of 1.56 inches.

2. Using a 60° threading tool cut a 0.062-inch deep V groove around the part, 0.375 inch back from the free end of the stock. The .375-inch dimension can be established by aligning the tip of the cutting

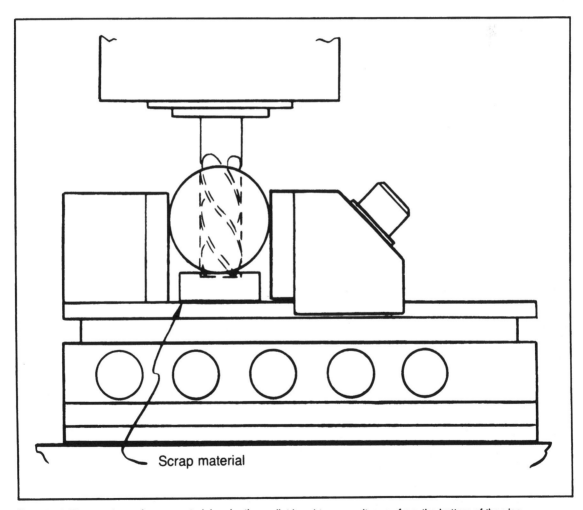

Scrap material

Fig. 11-15. Place a piece of scrap material under the mallet head to space it away from the bottom of the vise.

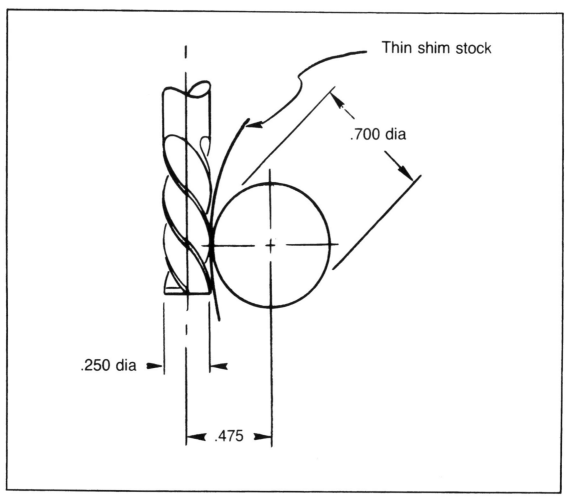

Fig. 11-16. To locate the cutter directly over the center of the mallet, bring the cutter up until it just touches the mallet. A thin piece of shim will prevent damaging the finish on the mallet. Once the edge of the part has been found, the distance to center can be found by adding the cutter diameter to the part diameter and dividing by two. The part can then be repositioned using the micrometer dial on the Y axis feedscrew to measure the distance.

tool with the end of the part, then using the feedscrew micrometer dial to move the tool over the 0.375-inch dimension. The depth of the groove is established in a similar manner. Bring the tool in until it just touches the part (Fig. 11-21). Then use the micrometer dial on the crossfeed feedscrew to bring the tool in an additional 0.062 inches.

3. The 0.12 inch radius groove can now be made with a ¼-inch diameter round file. The V notch will not only locate the file on the part, but it

will also indicate the proper depth of the radiused cut.

4. Use a fine flat file to cut a slight crown on the face of the hammer and to cut the 0.020-inch × 45° chamfer around the face. Following this, polish all of the surfaces with a fine emery. Don't forget to cover the slides on the lathe when polishing.

5. Remove the part from the lathe and rechuck it so that you can work the ball peen end. The

first step in cutting the ball peen will be to face the end and then reduce the diameter to 0.625 inches.

6. Use a 60° threading tool and cut a V groove .437 inches from the end of the stock 0.062 inches deep. Now, using a ¼-inch diameter file, cut the second 0.125-inch radiused groove.

7. Rotate the headstock of the lathe 45° and cut a 5/32-inch × 45° chamfer on the end of the part (Fig. 11-22).

8. Before actually cutting the radius on the ball peen, make yourself a template by cutting a 0.625-inch diameter circle out of a piece of card stock.

9. The spherically radiused ball peen is cut with a flat file. Take your time. Cut a little bit and check the result with the template, then cut a little bit more. When the peen has been cut to fit the template, polish it with fine emery cloth.

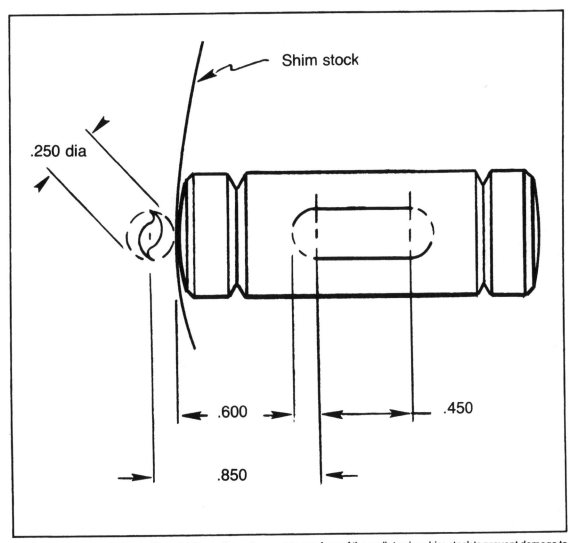

Fig. 11-17. To locate the start of the slot, locate the cutter next to one face of the mallet using shim stock to prevent damage to the finish on the mallet. Then use the micrometer dial on the X axis feedscrew to position the cutter.

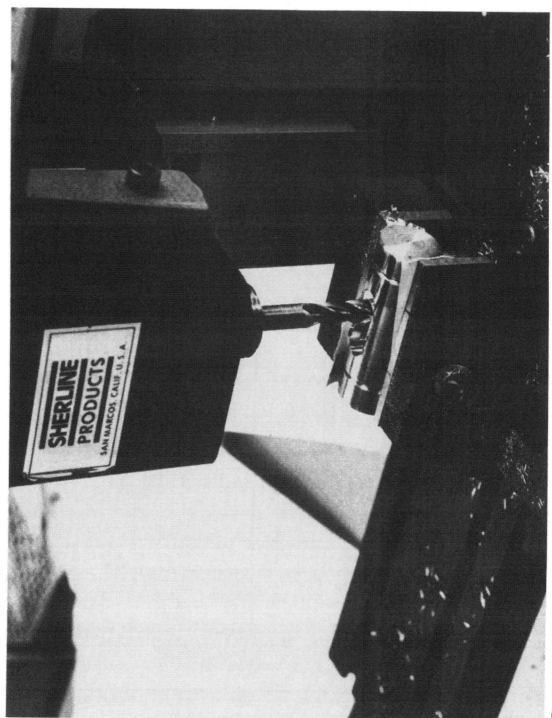

Fig. 11-18. Machining the slot in the mallet.

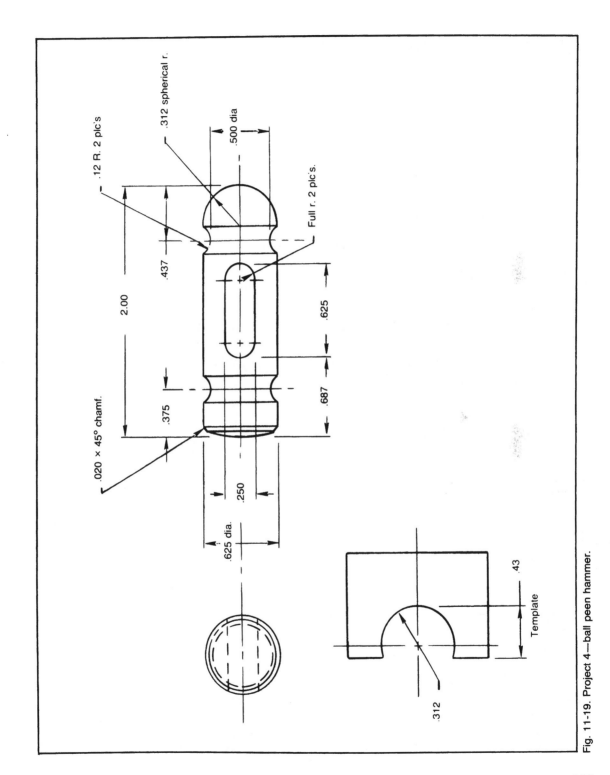

.12 R. 2 plc's

.312 spherical r.

.500 dia

.437

2.00

.625

.375

.687

Full r. 2 plc's.

.020 × 45° chamf.

.250

.625 dia.

.43

.312

Template

Fig. 11-19. Project 4—ball peen hammer.

229

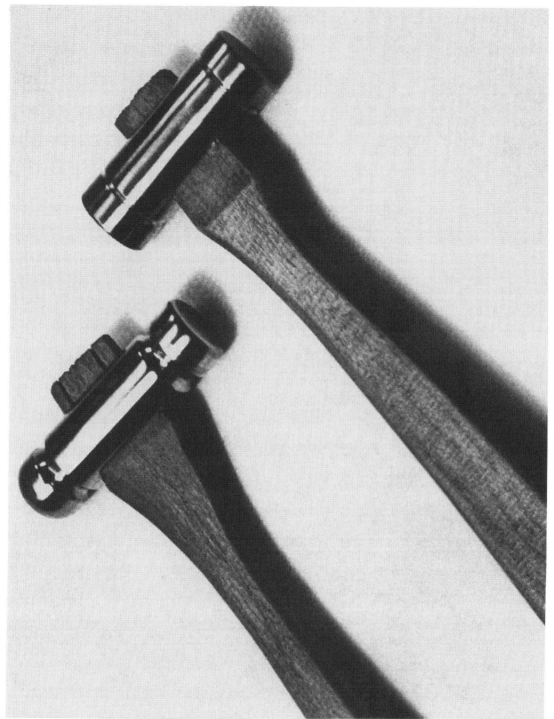

Fig. 11-20. A close-up shot of the finished ball peen hammer and brass mallet.

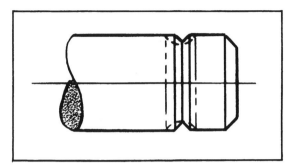

Fig. 11-21. The machined V groove as it looks prior to shaping the 0.12-inch radius groove.

The next operation is to cut the hole for the handle. This operation is done using a 0.250-inch diameter end mill ground for plunge cutting. The procedure is the same as outlined for the mallet.

When the slot is finished, the part can be removed from the mill and deburred. Heat treating should be done as outlined in the previous chapter. After hardening, the part should be tempered at 425° F. (light straw color). The hammer can then be given its final polishing and fitted with a wooden handle. As with the mallet, a handle can be purchased at most hardware stores or you can make your own from a piece of hardwood.

PROJECT 5: SETUP TOOL

This setup tool (Fig. 11-24) will be a big help to you when making quick setups on your milling machine. It will enable you to square a vise or a part to the bed of the machine quickly and easily with a fair degree of accuracy. It is used by locating the lip of the tool against the inside edge of a T slot or against the forward edge of the table. The side edge of the tool can then be used to square your part. The diagonal side can be used for making 30° or 60° setups.

The tool shown here has been designed for fabrication and use on the Sherline milling machine. If you are using a different type of machine, the dimensions may need to be changed. The short leg of the tool should be a little shorter than the width of the milling table. The three hold-down screw holes can then be located to the machine's T slots.

To make your setup tool, you will need a piece of ¼-inch aluminum plate. Alloy 6061 or 2024 in a T6 condition will do nicely. It should be saw cut to a triangular shape allowing at least 1/16 inch on all sides for finish cuts.

1. Locate the three .218-inch diameter holes allowing for the .062-inch excess material (0.300-inch + 0.062-inch = 0.362-inch; 0.780-inch + 0.062-inch = 0.842-inch). The stock can be clamped to the milling table and the mill used as a drill press for this operation, or you can lay out the holes, center punch them, and drill them using a hand-held drill motor. If you do the drilling in your milling machine, don't forget to place a piece of scrap material under the part so that you won't drill into the milling machine table. After drilling, countersink the holes so that the bottom of the countersink is even with the bottom of the hole.

2. Secure the plate to the bed of the milling machine using T nuts and flathead screws in each of the three holes. The base leg of the tool should be parallel to the back edge of the milling machine table. Use a piece of scrap aluminum under the part to space it up away from the milling machine table. This will give you clearance for milling around the outside of the part.

3. Using a 0.250-inch diameter cutter, cut the base leg of the triangle to obtain the 0.780-inch dimension. To make this cut, you will be using the X axis feedscrew, feeding from left to right. (Remember the difference between up and down mill-

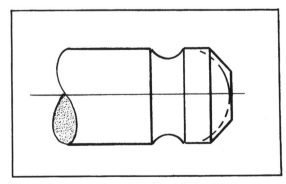

Fig. 11-22. The machined 5/32-inch × 45° chamfer prior to shaping the ball peen.

231

Fig. 11-23. Machining the 5/32-inch × 45° chamfer for the ball peen.

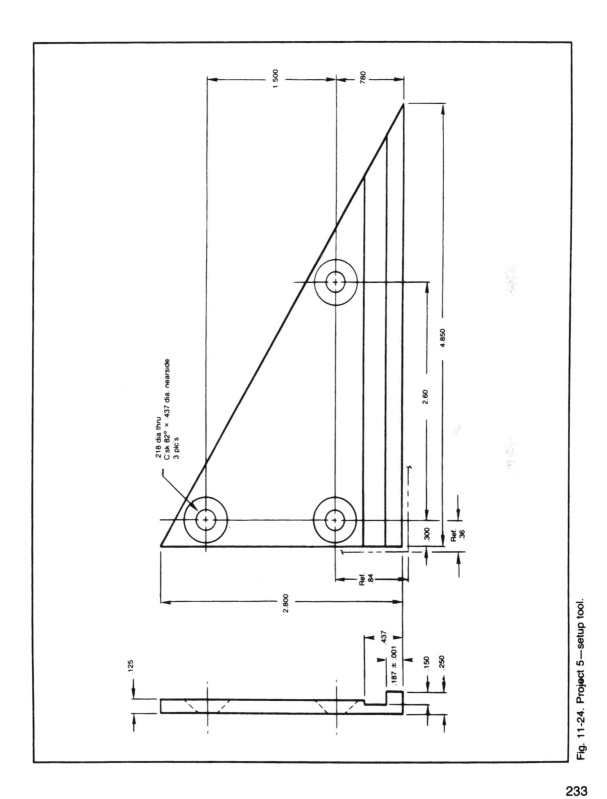

.218 dia thru
C'sk 82° × .437 dia. nearside
3 pic's

1.500
.780
4.850
2.60
.300
Ref. .36
Ref. .84
2.800

.125
.437
.187 ± .001
.150
.250

Fig. 11-24. Project 5—setup tool.

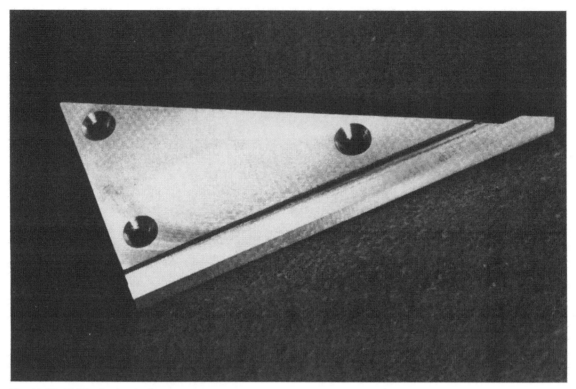

Fig. 11-25. Finished setup tool.

ing.) During all X axis cuts, lock the Y axis slides to prevent any unwanted movement. Likewise, on all Y axis cuts lock the X axis slide.

4. Cut a 0.250-inch wide × 0.150-inch deep slot parallel to the base leg holding the .187-inch dimension. This is done by using the Y axis feedscrew to move the part over exactly 0.187 inches, plus the cutter diameter (0.250-inch) or 0.437-inch. The slot should be cut using a series of cuts approximately 0.030 inch deep. Be sure to lock the Y axis slide during these cuts.

5. Using the Y axis feedscrew, move the part exactly 2.613 inches and cut off the tip of the triangle opposite the base leg. This will establish the length of the side leg at exactly 2.800 inches.

6. Cut the side leg of the tool holding the 0.300-inch dimension, then using the X axis feedscrew, move the part exactly 5.100 inches (calculated 4.850 inches + cutter diameter), and cut off

the tip of the triangle opposite the side leg. This will establish the length of the base leg of 4.850 inches.

7. Change the cutter to a fly cutter set for cutting a 1½-inch diameter. The cuts are made feeding left to right, taking a depth of approximately 0.020 inches with each pass (Fig. 11-26).

8. Replace the fly cutter with the ¼-inch diameter end mill. Reclamp the part with the point of the 30° angle and the point of the 60° angle directly over the edge of the second T slot (Fig. 11-27). A small 30° drafting triangle held against the base leg of the tool and a dial indicator can be used to verify the accuracy of your setup. When the setup is complete, cut the hypotenuse of the tool.

9. Deburr the tool, using some fine emery cloth or paper, and it is ready for use (Fig. 11-28).

PROJECT 6: PARALLEL BARS

You will probably want to make parallel bars

Fig. 11-26. Fly cutting the setup tool.

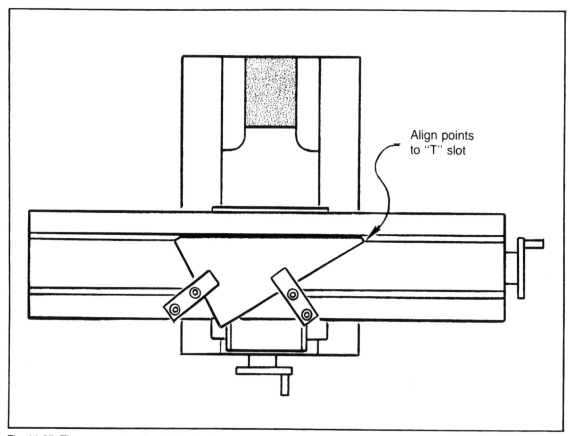

Fig. 11-27. The proper setup for milling the hypotonuse of the setup tool.

Align points to "T" slot

(Fig. 11-29) in several different sizes. All of mine are 3.00 inches long, and I have made them with the width equal to half the height: ¼ × ½-inch, ⅜ × ¾-inch, and ½ × 1-inch. The bars are slightly smaller than these dimensions because I started with bar stock for these sizes and used minimal cleanup cuts to true them up.

Commercially-made bars are generally of high-carbon, hardened steel. For use in the small shop, I have found that bars made from low-carbon steel or even aluminum are just as good. They are a lot easier to make, and given a reasonable amount of care they will last a long time. To make yours, you will need some bar stock of the same approximate size that you want the bars to be. You will want to make two bars at the same time for a matched pair.

Use your milling vise to hold the parts. It should be set up with the jaws parallel to the X axis of the milling machine. You can use your setup tool (project 5) to square the vise.

1. To start the job, clamp the bars side by side in the vise with side 1 up (Fig. 11-30). Use a piece of scrap aluminum on either side to minimize the chance of accidentally cutting into the vise jaws.

2. The cuts are made with a fly cutter. Use minimal cuts, just deep enough to clean up and true the surfaces. When side 1 is complete, reclamp the parts in the vise with side 2 up. Use a dial indicator clamped to the milling head and indicating on side 1 of the bars to determine that both parts are parallel to each other, and that side 1 will be parallel to side 2 when side 2 is completed (Fig. 11-32). When the

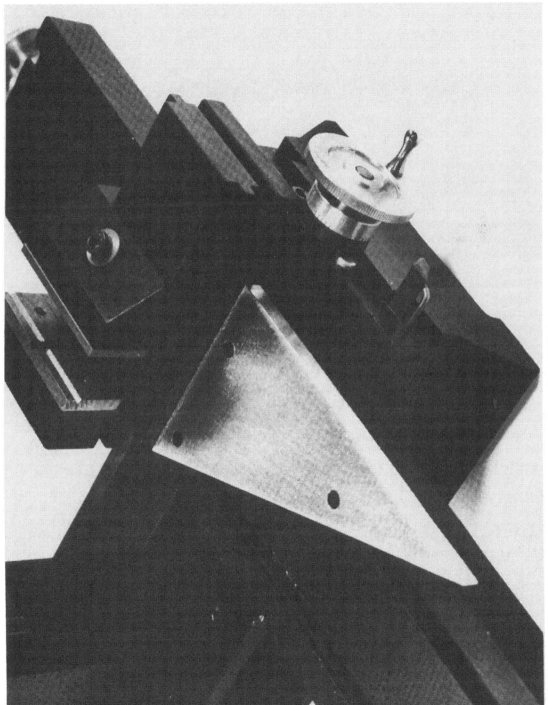

Fig. 11-28. Using the finished setup tool to square a vise to the milling machine.

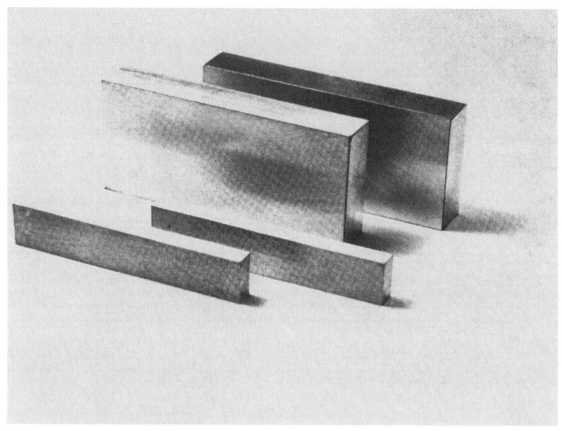

Fig. 11-29. Two sets of finished parallel bars.

parts have been indicated in to your satisfaction, remove the dial indicator and cut side 2.

3. Repeat the above operations for sides 3 and 4. The last milling operation should be to true up and square the ends of the bars. After the parts are completely machined, they should be deburred with some fine emery paper. The resulting parallels may not be as accurate as a pair of commercially ground parallels, but they should work fine for most applications.

PROJECT 7: MACHINIST'S CLAMP

Making a machinist's clamp (Figs. 11-33 and 11-34) requires the use of both a lathe and a milling machine. The type of material used is not impor-

tant. Any kind of steel will do, but the use of a free-machining steel will make the job easier. The screws are made from 7/16-inch hex bar stock. The clamp jaws are made from ½-inch square bar stock. The retaining clip is made from a piece of 1/16-inch thick sheet stock.

1. The first step in making the clamp is to cut the raw stock to length. Note that the two clamping screws are different. One is 3.10 inches long and the other is 3.25 inches long. In machining the two screws you will want to use your three-jaw chuck and support the free end of the material with a center.

2. Turn the material down to the threading size (0.248 ± .001-inch diameter). The actual

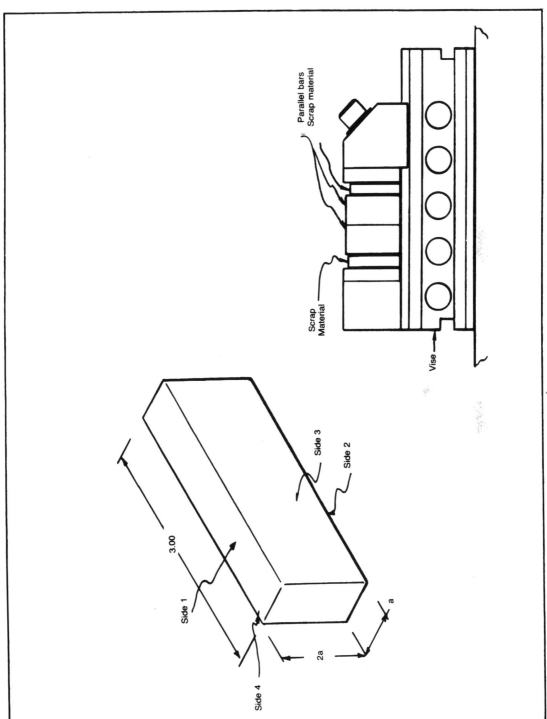

Fig. 11-30. Use scrap material on each side of the bars to provide clearance so you won't cut into the vise jaws.

Fig. 11-31. Fly cutting the first side of the parallel bars.

Fig. 11-32. Indicating the first side of the parallel bars prior to cutting the second side.

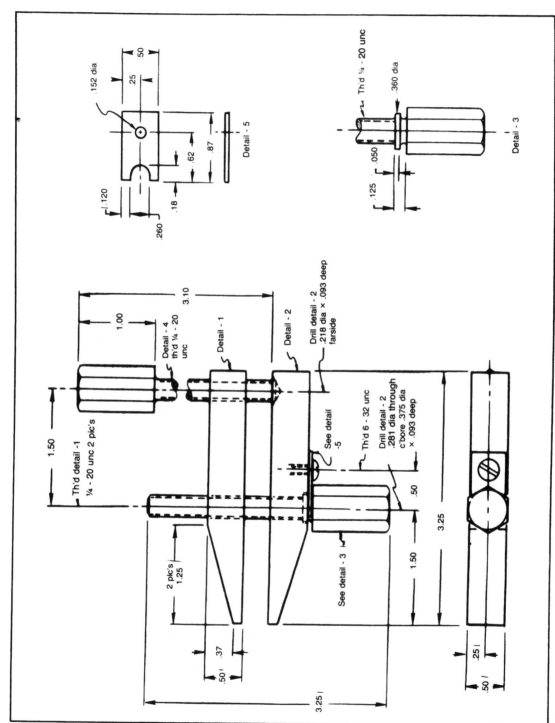

Fig. 11-33. Project 7—machinist's clamp.

242

Fig. 11-34. Finished machinist's clamp.

threading can be done with your lathe's threading attachment or with a ¼-20 UNC die. After the parts are threaded, they should be deburred and polished.

When preparing to make the clamp jaws, note that they are also different. One is threaded for the clamping screws; the other is not. To machine these parts follow the same steps that you used in machining your parallel bars, squaring the four sides and the ends. Parallelism of the four sides is not critical, so the steps involving indicating the part using a dial indicator may be eliminated.

3. The 0.37 × 1.25-inch chamfer is machined by clamping the parts at an angle in the vise. Prior to clamping them, scribe a line indicating the chamfer on the side of one part. Both parts can then be clamped in the vise side by side with the scribe line parallel to the top of the vise.

4. All of the holes can be drilled in the milling machine. Use the micrometer scales on the feedscrews in the same way that you used them to locate the slot in the mallet head. The threads are cut using standard taps. Make sure that the tap enters the part squarely. Any misalignment may cause the screws to bind. The two jaws can be clamped together, and the clearance holes in the one can be used as a guide to align the tap squarely.

5. The retaining clip is made by roughly sawing the part out and then filing to shape. When all of

the parts have been machined, deburred, and polished, they may be assembled. Like all items made from steel, these clamps will rust if not cared for properly. A light coat of oil will keep them looking new for a long time. For a more lasting finish, try the black, baked oil finish outlined in Chapter 10.

PROJECT 8: ANGLE PLATE

Any thick-walled piece of angle iron can be used to make a small angle plate (Figs. 11-35 and 11-36). The piece I used in the accompanying illus-

trations was 2.00 × 0.37 × 2.00 inches long. Before starting any machining, the material should be annealed to relieve any internal stresses.

To make an angle plate you will need to rotate the milling head so that the spindle axis is exactly parallel to the bed of the machine, and this presents a problem. With the milling head rotated, the minimum height of the spindle is too great to permit machining the angle. To get around this problem, you will need to make a large spacer (Fig. 11-37) to support the part. To make the spacer block, I found a block of aluminum measuring 2⅛ × 2½ × 2⅞

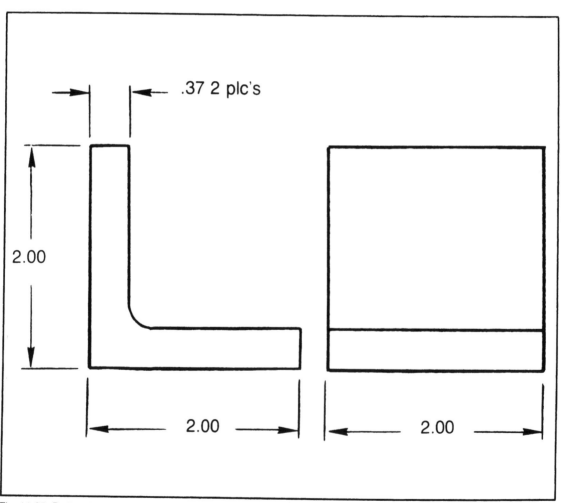

Fig. 11-35. Project 8—angle plate.

Fig. 11-36. Finished angle plate.

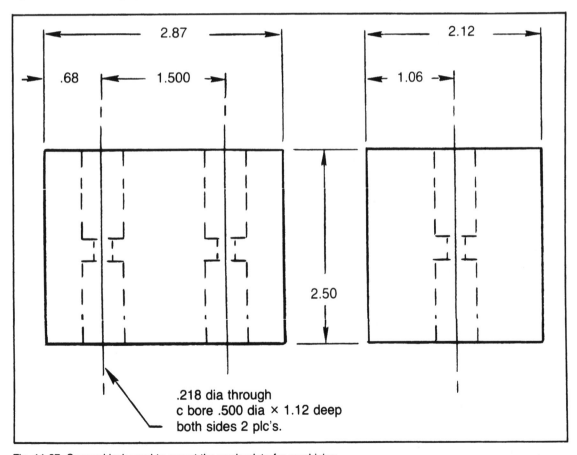

.218 dia through
c bore .500 dia × 1.12 deep
both sides 2 plc's.

Fig. 11-37. Spacer block used to mount the angle plate for machining.

245

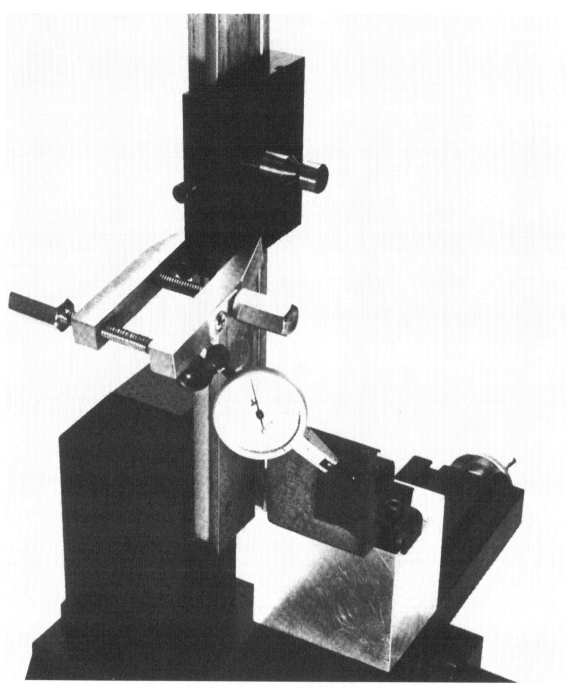

Fig. 11-38. Indicating the first leg of the angle prior to machining. This step helps to minimize the amount of material that has to be removed. Note the use of the machinist's clamp to hold the dial indicator.

inches. I drilled and counterbored the block and mounted it to the milling table with a pair of hold-down screws. After mounting the block to the table, I took a minimal cut with a fly cutter to machine the top of the block parallel to the table top. I also drilled and tapped three 10-32 holes in the top of the block for clamping the angle to the spacer.

When the spacer block is completed, you are ready to set up the milling machine and start work. First, rotate the milling head so that the spindle axis is exactly parallel to the top surface of the spacer block. To do this, secure a dial indicator to the spindle and indicate on the vertical leg of a square clamped firmly to the spacer block. With this operation complete, mount your part on the spacer block.

1. The angle should be clamped to the spacer block so that the vertical leg is parallel to the Y axis of the machine. The face of the vertical leg should protrude past the edge of the spacer block by approximately 1/8 inch. This will preclude the possibility of cutting into the spacer block as you machine the angle (Fig. 11-38).

2. A fly cutter is used to machine the face of the angle. Mount the cutter into the machine and lower the milling head as low as it will go. The cutting tool can then be adjusted to cut a circle that just clears the two bottom corners of the angle (Fig. 11-39).

3. Raise the milling head so that the cutter just clears the top of the angle, and you are ready to start machining. You will be feeding the cutter down from top to bottom. The depth of cut should be approximately 0.005 inch, and you may need several cuts to completely machine the surface (Fig. 11-40). Remove just enough material to clean up and true the entire surface.

4. When the face cut is complete, remove the fly cutter and replace it with a 0.250-inch diameter end mill. The end mill is used to true up the upper end of the angle (Fig. 11-41).

The same procedure is then repeated for the second leg of the angle. When the second leg has been finished, you will be ready to square up the sides. The sides are cut with a fly cutter, and the part is held in a vise (Fig. 11-42). To set the milling machine up, rotate the milling head back to its vertical position. Then, using your setup tool (project 5), square the vise with the jaws parallel to the X axis of the milling machine and bolt it down. The part is clamped in the vise on its side. A square can be used to help position it so that the side will be perpendicular to the other machined surfaces.

After both sides have been machined, the part should be deburred. It is now ready for use. As with all steel tools, it should always be coated with a thin coat of oil to prevent rust.

PROJECT 9: SINE BAR

Alloy 4130 steel is a good choice of material for making your sine bar (Figs. 11-43 and 11-44). You will need a piece of 9/16-inch diameter × 3.50-inch long round bar stock and a piece of 0.50 × 1.00 × 6.00-inch bar stock.

1. To start the job, mount the round bar stock in your lathe using a three-jaw chuck. Next, face and center drill the free end of the stock. Using a center to support the free end of the bar, machine the outside diameter for a distance of 3.12 inches to a diameter of 0.500 inches, ± .001. Check the diame-

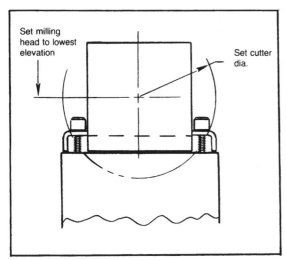

Set milling head to lowest elevation

Set cutter dia.

Fig. 11-39. Adjusting the proper swing for the fly cutter.

Fig. 11-40. Fly cutting the first leg.

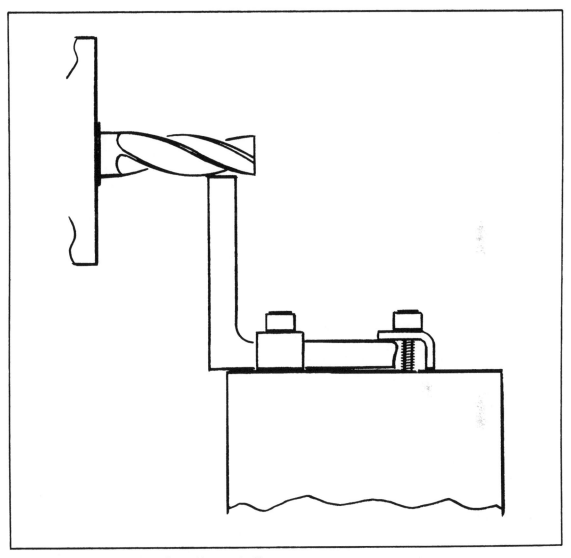

Fig. 11-41. Machining the end surface with an end mill.

ter in several places along the bar. If the tailstock is not in perfect alignment with the headstock, you may end up with a slight taper in your workpiece. If this is the case, cut the smallest diameter to size and then use a fine file to work the rest of the bar to size. When the correct diameter has been obtained, polish the surface and use a parting tool to cut off two 1.50-inch long sections. These sections can then be drilled and tapped for a 10-32 screw.

2. To machine the bar, first square up each of the six sides using a fly cutter and an end mill. The procedure for squaring up the bar is the same as used in machining the parallel bars. It is not necessary to strive for any great degree of accuracy in regard to the parallelism of the sides at this time. When all of the sides are squared up, machine the two 90° V grooves.

3. This operation is accomplished with a

Fig. 11-42. Fly cutting the sides of the angle plate.

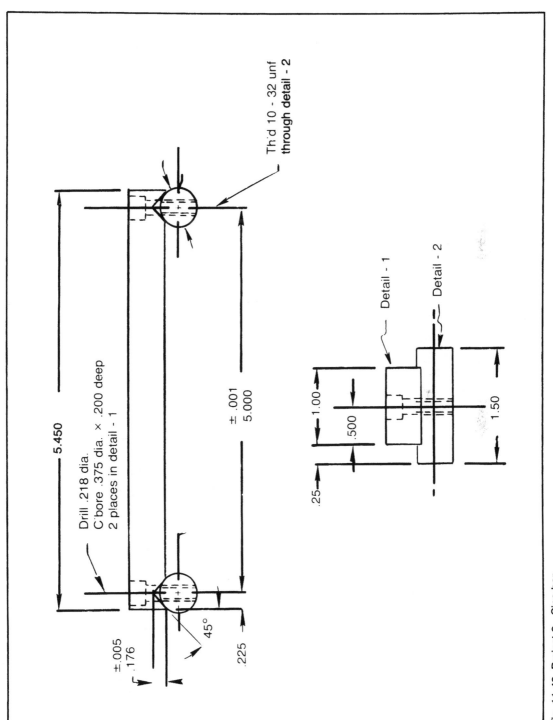

Fig. 11-43. Project 9—Sine bar.

Fig. 11-44. Finished sine bar.

0.250-inch diameter end mill, with the headstock rotated to an angle of 45°. The distance between the two V grooves is critical and should be established using the micrometer indicator on the feedscrew. The depth of the V notch is also somewhat critical and care should be taken to make both grooves the same.

4. The next step is to drill and counterbore the two screw holes in the bar (Fig. 11-45). When

this has been accomplished, deburr the bar and assemble the two round bars onto it.

5. Clamp the sine bar to the bed of your milling machine, using four strap clamps, one on each end of the round bars protruding past the side of the sine bar. Then, using a fly cutter, take a final cut on the upper surface of the sine bar. This will compensate for any inaccuracy in the diameter of the round bars or depth of the V notches.

Fig. 11-45. Drilling the ends of the sine bar in a milling machine. Note the use of the machinist's jack to help support the end of the bar.

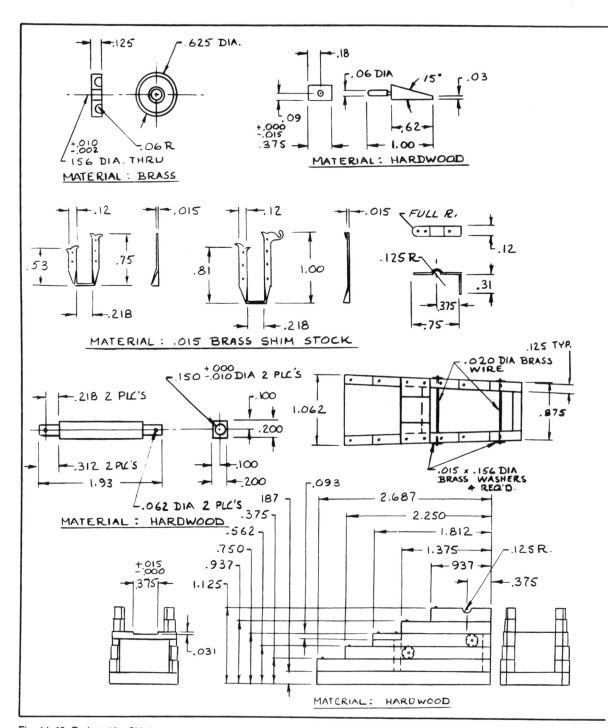

Fig. 11-46. Project 10—Ship's cannon.

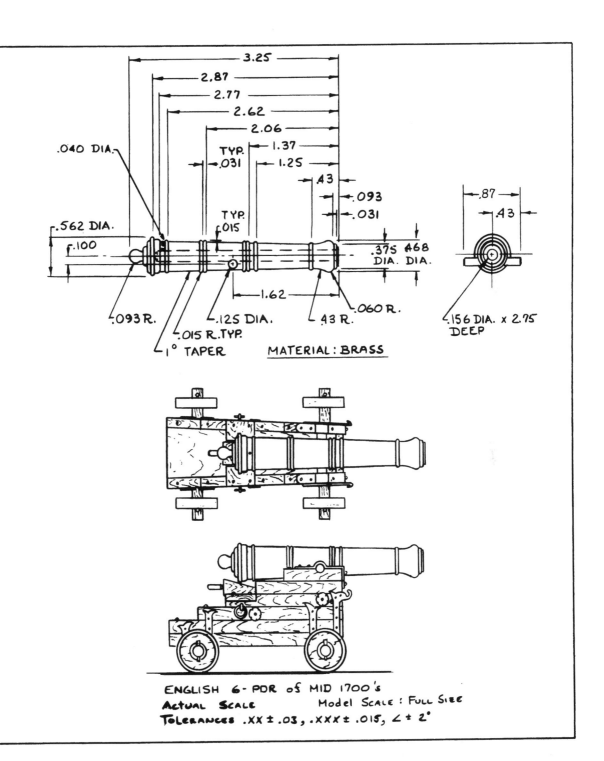

3.25
2.87
2.77
2.62
2.06
TYP. — 1.37
.031
1.25
.43
.093
.031

.040 DIA.

TYP.
.015

.562 DIA.

.100

.375 .468
DIA. DIA.

.87
.43

.093 R.

.125 DIA.

.015 R. TYP.

1° TAPER

1.62

.43 R.

.060 R.

MATERIAL: BRASS

.156 DIA. x 2.75
DEEP

ENGLISH 6-PDR of MID 1700's
ACTUAL SCALE   Model Scale: Full Size
Tolerances .XX ± .03, .XXX ± .015, ∠ ± 2°

255

To use your sine bar, simply look up in a sine chart for the sine of any angle you may need. Then divide the sine by 2 and move the decimal over one place to the right. This will be the size spacer you need under one end of the bar to tilt it to the required angle.

PROJECT 10: SHIP'S CANNON

This model is typical of the style of cannon used by the English Navy in the mid 1700s. The barrels were made of brass, as were most of the metal fittings. The carriages were made of hardwood, usually oak. The projectile was a 6-pound iron ball (Figs. 11-46 and 11-47). To make the barrel of your cannon, you will need a piece of brass rod ⅝-inch diameter × 3.50-inch long.

1. The first machining operation will be to drill a 0.125-inch diameter hole for the pivot pin. This operation is best accomplished in your milling machine or in a lathe equipped with a vertical milling attachment. Use the techniques discussed in earlier projects to locate the hole 0.100 inch from the center of the barrel and 1.62 inch + 0.03 inch from the muzzle end of the barrel. The extra 0.03 inch will be machined away when you face the end of the barrel. Start the hole with a center drill. Then drill it out slightly undersize. Finally enlarge it to size using a 0.125-inch diameter reamer or drill bit.

2. Drill the touch hole (Fig. 11-48). The touch hole is the 0.040-inch diameter hole that intersects the bore. It is located exactly on center, 2.710 inches back from the muzzle end of the barrel. To make sure that the touch hole is at 90° to the pivot hole, stick a 0.125-inch diameter drill shank through the pivot hole and position the barrel in your vise with the drill resting on the top of the vise. To clamp the barrel in the vise, you will need to angle the breach end downward approximately 15°. The actual angle is not important, but don't slope it any steeper than you have to. This hole should also be started with a center drill.

3. The next step is boring the barrel. This operation is accomplished in your lathe. Mount the part using a three-jaw chuck. The muzzle end of the

Fig. 11-47. Finished ship's cannon.

barrel will be facing the tailstock. A center drill is used to start the bore. Drill the bore using a 9/64-inch (0.140) diameter drill. Put a band of masking tape around the drill bit to indicate when you have reached the proper depth.

Drilling the barrel should be done slowly, backing the drill completely out of the hole frequently to clear the chips. This becomes increasingly important as you drill deeper into the part. It is also important not to build up too much heat as the thermal expansion will cause the drill to bind up in the hole. After the hole has been drilled, ream it to size using a 5/32-inch (0.156) diameter reamer or drill bit.

4. Face the muzzle end of the barrel. Then machine the 0.375-inch diameter × 0.093-inch and 0.312-inch diameter × 0.031-inch steps just behind the muzzle. Then turn the muzzle end of the barrel down to 0.468-inch diameter for a distance of 0.43 inches.

5. The next step requires rotating the headstock 1° to cut the 1° taper. The protractor scale on the headstock is not accurate enough to make this

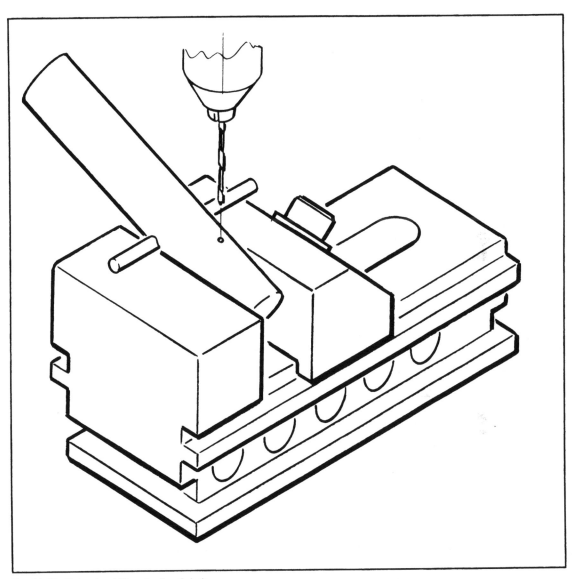

Fig. 11-48. Setup for drilling the touch hole.

setting, so you need a method of measuring the exact angle. This is not as hard as it might seem. Using a trigonometry table, you can calculate over some distance along the length of the barrel how much reduction there will be in the radius of the barrel. To do this, visualize an imaginary triangle placed alongside the barrel (Fig. 11-49). Side "b" is parallel to the center line of the barrel. Side "c" is

the outside of the barrel, and side "a" is the distance between the barrel and the free end of side "b." Angle A is 1° as shown on the print.

Looking at the formula shown in the Appendix, you will find that "a" is equal to "b" times the tangent of angle A. Using a trigonometric table, you will find that the tangent of 1° is equal to 0.0175. For convenience you can say that side "b" is 2.00 inches

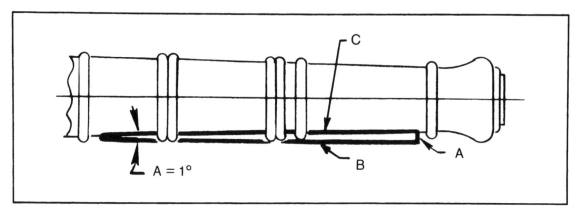

Fig. 11-49. Calculating the offset for machining the 1° taper.

long (the longest even measurement you can make along the side of the barrel). Multiplying 2.00 inches times 0.0175, you get 0.0349 inches. Rounding this number off you will have "a" = 0.035 inches. You can now use the micrometer scales on your lathe's feedscrews to measure the 0.035 inches and the 2.00 inches. This method of estab-

lishing a taper is quite accurate and can be used in a number of different applications (Fig. 11-50).

6. Once you have the headstock set up for machining the 1° taper, you can start rough machining the taper in the area between 0.30 inches and 2.87 inches from the muzzle end of the barrel. You want to continue roughing the taper until the

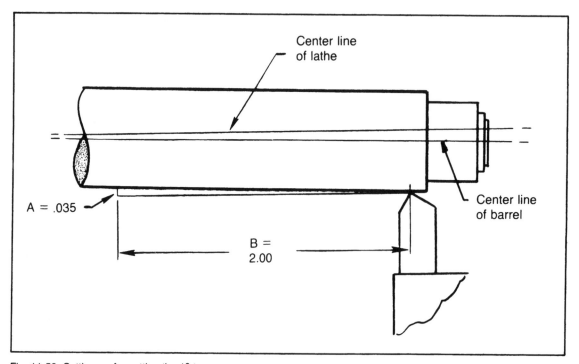

Fig. 11-50. Setting up for cutting the 1° taper.

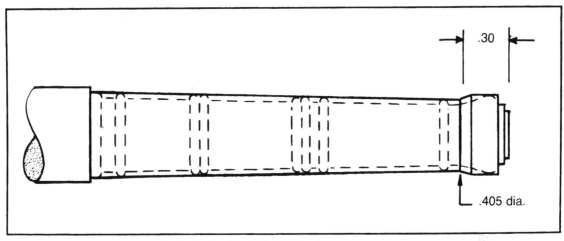

Fig. 11-51. Rough-machined barrel allowing for the reinforcing bands.

smaller diameter (0.30 inches from the end of the barrel) is reduced to a diameter of 0.405 inches (Fig. 11-51). This leaves enough material for the 0.015-inch high reinforcing bands around the barrel (calculate: 0.375-inch diameter + 0.015 inch + 0.015 inch). The taper can now be finish-cut to size, leaving material for the reinforcing bands. See Fig. 11-52. Final shaping of the bands and radiuses (0.43-inch radius and 0.060-inch radius) at the muzzle are done with a small, fine file. Next, the entire machined portion of the barrel can be polished.

7. To turn the breech end of the barrel, you will have to mount the barrel on a temporary arbor (Fig. 11-53). The arbor can be made from a piece of scrap bar stock, preferably steel. To make the arbor, turn it down to a diameter that just slips into the cannon's bore (0.156-inch + .000-inch, −

.001-inch diameter) for a distance of 1.80 inches. Then file a small notch near the free end of the arbor so that when the barrel is slid over the arbor, a 0.125-inch diameter pin can be inserted in the pivot hole to lock the barrel in place. The pin can be held in place by wrapping masking tape around the barrel. Make sure the pin is secure enough so that it won't fly out while you are machining.

The breech end of the barrel is machined following the same techniques outlined for the rest of the barrel. Final shaping is accomplished with a file. Polishing should also be completed before removing the arbor. When the machining and polishing are all finished, you can remove the arbor and press fit a 0.125-inch diameter × 0.87-inch long brass pivot pin in place. The pivot pin intersects the bore, so if you want your project to be perfect, you may choose

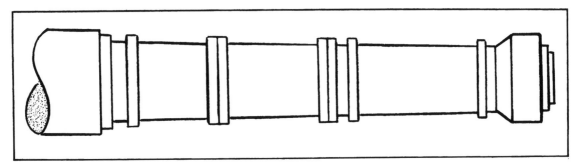

Fig. 11-52. The machined barrel prior to shaping with a file.

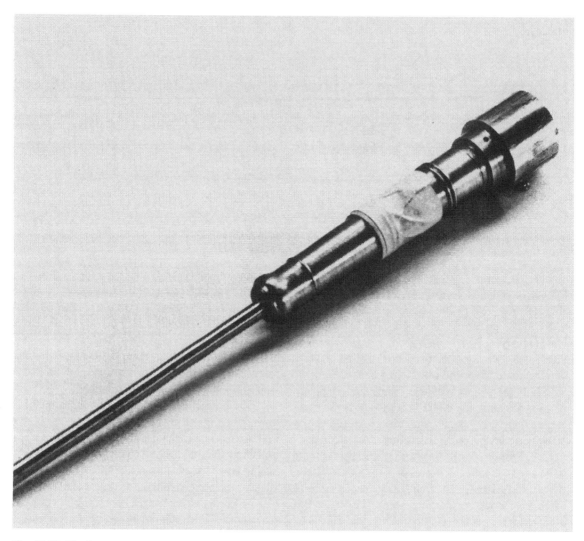

Fig. 11-53. The barrel mounted on a temporary arbor so the breech can be machined.

to run a reamer into the barrel to open up the bore. Since this cannon is intended as a nonfiring decoration only, this step is not necessary.

The carriage can be made of any kind of wood, using standard woodworking procedures. Finish it with a dark stain. The blueprint calls for the wheels to be made of brass, but wood is satisfactory. To make the brass straps, you will need to clamp the brass shim stock between two pieces of 0.062-inch aluminum and then machine and file to shape. The aluminum provides enough stiffness to enable a good job of cutting the thinner shim stock. To secure the straps in place, I machined some small brass nails from a piece of brazing rod. I'm sure that small tacks could have been purchased from a hobby shop.

Appendix
Tables

Metric/Decimal Equivalents.

MM	Decimal		MM	Decimal
0.1	0.003937		10.0	0.393701
.2	.007874		20.0	.787402
.3	.011811		30.0	1.181103
.4	.015748		40.0	1.574804
.5	.019685		50.0	1.968505
.6	.023622		60.0	2.362206
.7	.027559		70.0	2.755907
.8	.031496		80.0	3.149608
.9	.035430		90.0	3.543309
1.0	.039370		100.0	3.93701
2.0	.078740		200.0	7.87402
3.0	.118110		300.0	11.81103
4.0	.157480		400.0	15.74804
5.0	.196850		500.0	19.68505
6.0	.236220		600.0	23.62206
7.0	.275591		700.0	27.55907
8.0	.314961		800.0	31.49608
9.0	.354331		900.0	35.43309

Fraction/Decimal/Metric Equivalents.

Fraction	Decimal	MM	Fraction	Decimal	MM
1/64	0.015625	0.3969	33/64	0.15625	13.0969
1/32	.03125	.7938	17/32	.53125	13.4938
3/64	.046875	1.1906	35/64	.546875	13.8906
1/16	.0625	1.5875	9/16	.5625	14.2875
5/64	.078125	1.9844	37/64	.578125	14.6844
3/32	.09375	2.3813	19/32	.59375	15.0813
7/64	.109375	2.7781	39/64	.609375	15.4781
1/8	.1250	3.1750	5/8	.6250	15.8750
9/64	.140625	3.5719	41/64	.640625	16.2719
5/32	.15625	3.9688	21/32	.65625	16.6688
11/64	.171875	4.3656	43/64	.671875	17.0656
3/16	.1875	4.7625	11/16	.6875	17.4625
13/64	.203125	5.1594	45/64	.703125	17.8594
7/32	.21875	5.5563	23/32	.71875	18.2564
15/64	.234375	5.9531	47/64	.734375	18.6531
1/4	.2500	6.3500	3/4	.7500	19.0500
17/64	.265625	6.7469	49/64	.765625	19.4469
9/32	.28125	7.1438	25/32	.78125	19.8438
19/64	.296875	7.5406	51/64	.796875	20.2406
5/16	.3125	7.9375	13/16	.8125	20.6375
21/64	.328125	8.3344	53/64	.828125	21.0344
11/32	.34375	8.7313	27/32	.84375	21.4313
23/64	.359375	9.1281	55/64	.859375	21.8281
3/8	.3750	9.5250	7/8	.8750	22.2250
25/64	.390625	9.9219	57/64	.890625	22.6219
13/32	.40625	10.3188	29/32	.90625	23.0188
27/64	.421875	10.7156	59/64	.921876	23.4156
7/16	.4375	11.1125	15/16	.9375	23.8125
29/64	.453125	11.5094	61/64	.953125	24.2094
15/32	.46875	11.9063	31/32	.96875	24.6063
31/64	.484375	12.3031	63/64	.984375	25.0031
1/2	.5000	12.7000	1	1.0000	25.4000

Standard Drill Sizes.

Fraction	Gauge	Decimal
	80	0.0135
	79	.0145
1/64		.015625
	78	.016
	77	.018
	76	.020
	75	.021
	74	.0225
	73	.024
	72	.025
	71	.026
	70	.028
	69	.02925
	68	.031
1/32		.03125
	67	.032
	66	.033
	65	.035
	64	.036
	63	.037
	62	.038
	61	0.39
	60	.040
	59	.041
	58	.042
	57	.043
	56	.0465
3/64		.046875
	55	.052
	54	.055
	53	.0595
1/16		.0625
	52	.0635
	51	.067

Fraction	Gauge	Decimal
	50	0.070
	49	.073
	48	.076
5/64		.078125
	47	.0785
	46	.081
	45	.082
	44	.086
	43	.089
	42	.0935
	41	.096
	40	.098
	39	.0995
	38	.1015
	37	.104
	36	.1065
7/64		.109375
	35	.110
	34	.111
	33	.113
	32	.116
	31	.120
1/8		.1250
	30	.1285
	29	.1360
	28	.1405
9/64		.140625
	27	.144
	26	.147
	25	.1495
	24	.152
	23	.154
5/32		.15625
	22	.157

Fraction	Gauge	Decimal
	21	0.159
	20	.161
	19	.166
	18	.1695
11/64		.171875
	17	.173
	16	.177
	15	.180
	14	.182
	13	.185
3/16		.1875
	12	.189
	11	.191
	10	.1935
	9	.196
	8	.199
	7	.201
	6	.204
	5	.2055
	4	.209
	3	.213
7/32		.21875
	2	.221
	1	.228
	A	.234
15/64		.234375
	B	.238
	C	.242
	D	.246
	E	.250
1/4		.250
	F	.257
	G	.261
17/64		.265625
	H	.266

Fraction	Gauge	Decimal
	I	0.272
	J	.277
	K	.281
9/32		.28125
	L	.290
	M	.295
19/64		.296875
	N	.302
5/16.		.3125
	O	.316
	P	.323
21/64		.328125
	Q	.332
	R	.339
11/32		.34375
	S	.348
	T	.358
23/64		.359375
	U	.368
3/8		.375
	V	.377
	W	.386
25/64		.390625
	X	.397
	Y	.404
13/32		.40625
	Z	.413
27/64		.421825
7/16		.4375
29/64		.453125
15/32		.46875
31/64		.484375
1/2	•	.500

Standard Metric Drill Sizes.

MM	Decimal of Inch		MM	Decimal of Inch		MM	Decimal of Inch		MM	Decimal of Inch
0.4	0.015748		2.15	0.084645		4.75	0.187007		7.6	0.299212
0.5	0.019685		2.2	0.086614		4.8	0.188976		7.7	0.303149
0.55	0.021653		2.25	0.088582		4.9	0.192913		7.75	0.305117
0.6	0.023622		2.3	0.090551		5.0	0.196850		7.8	0.307086
0.65	0.02559		2.35	0.092519		5.1	0.200787		7.9	0.311023
0.7	0.027559		2.4	0.094488		5.2	0.204724		8.0	0.314960
0.75	0.029527		2.45	0.096456		5.25	0.206692		8.1	0.318397
0.8	0.031496		2.5	0.098425		5.3	0.208661		8.2	0.322834
0.85	0.033464		2.6	0.102362		5.4	0.212598		8.25	0.324802
0.9	0.035433		2.7	0.106299		5.5	0.216535		8.3	0.326771
0.95	0.037401		2.75	0.108267		5.6	0.220472		8.4	0.330708
1.0	0.039370		2.8	0.109236		5.7	0.224409		8.5	0.334645
1.05	0.041338		2.9	0.114173		5.75	0.226377		8.6	0.338582
1.1	0.043307		3.0	0.118110		5.8	0.228346		8.7	0.342519
1.15	0.045275		3.1	0.122047		5.9	0.232283		8.75	0.344487
1.2	0.047244		3.2	0.125984		6.0	0.236220		8.8	0.346356
1.25	0.049212		3.25	0.127952		6.1	0.240157		8.9	0.350393
1.3	0.051181		3.3	0.129921		6.2	0.244094		9.0	0.354330
1.35	0.053149		3.4	0.133858		6.25	0.246062		9.1	0.358267
1.4	0.055118		3.5	0.137795		6.3	0.248031		9.2	0.362204
1.45	0.057086		3.6	0.141732		6.4	0.251968		9.25	0.364172
1.5	0.059055		3.7	0.145669		6.5	0.255905		9.3	0.366141
1.55	0.061023		3.75	0.147637		6.6	0.259842		9.4	0.370078
1.6	0.062992		3.8	0.149606		6.7	0.263779		9.5	0.374015
1.65	0.064960		3.9	0.153543		6.75	0.265747		9.6	0.377952
1.7	0.066029		4.0	0.157480		6.8	0.267716		9.7	0.381889
1.75	0.068897		4.1	0.161417		6.9	0.271653		9.75	0.383857
1.8	0.070866		4.2	0.165354		7.0	0.275590		9.8	0.385826
1.85	0.072834		4.25	0.168322		7.1	0.279527		9.9	0.389763
1.9	0.074803		4.3	0.169291		7.2	0.283464		10.0	0.393700
1.95	0.076771		4.4	0.173228		7.25	0.285432		10.5	0.413385
2.0	0.078740		4.5	0.177165		7.3	0.287401		11.0	0.433070
2.05	0.080708		4.6	0.181102		7.4	0.291338		11.5	0.452754
2.1	0.082677		4.7	0.185039		7.5	0.295275		12.0	0.472440

Standard Gauges for Wire and Sheet Metal.

Gauge	Music Wire[1]	Brown & Sharp[2]	U.S. Standard[3]
0000	0.006	0.460	0.40625
000	.007	.410	.37500
00	.008	.365	.34375
0	.009	.325	.31250
1	.010	.289	.28125
2	.011	.258	.26562
3	.012	.229	.23910
4	.013	.204	.22420
5	.014	.182	.20920
6	.016	.162	.19430
7	.018	.144	.17930
8	.020	.128	.16440
9	.022	.114	.14950
10	.024	.102	.13450
11	.026	.091	.11960
12	.029	.081	.10460
13	.031	.072	.08970
14	.033	.064	.07470
15	.035	.057	.06730
16	.037	.051	.05980
17	.039	.045	.05380
18	.041	.040	.04780
19	.043	.036	.04180
20	.045	.032	.03590
21	.047	.028	.03290
22	.049	.025	.02990
23	.051	.023	.02690
24	.055	.020	.02390
25	.059	.018	.02090
26	.063	.016	.01790
27	.067	.014	.01640
28	.071	.0126	.01490
29	.075	.011	.01350
30	.080	.010	.01200
31	.085	.009	.01094
32	.090	.008	.01016
33	.095	.007	.00938
34	.100	.0063	.00859
35	.106	.0056	.00781
36	.112	.005	.00703

[1] Spring steel wire is usually specified in Music Wire Gauge.
[2] Brown & Sharp wire gauges are usually used for all wire sizes except spring steel.
[3] U.S. standard gauges are usually used to specify sheet metal thicknesses.

Threaded Fastener Information.

Thread Size	Thread Diameter	Threads per inch (A)	Tap Drill (B)		Depth		Clearance Drill (E)		Fillet Head	
			No.	Dec.	(C) Drill	(D) Tap	No.	Dec.	D	H
0	0.060	80	3/64	0.048	3/16	1/8	50	0.070	0.096	0.056
1	.073	64	53	.059	7/32	5/32	43	.089	.118	.071
		72	53	.059						
2	.086	56	50	.070	9/32	3/16	37	.104	.140	.083
		64	50	.070						
3	.099	48	5/64	.078	5/16	7/32	31	.120	.161	.095
		56	45	.082						
4	.112	40	43	.089	11/32	1/4	9/64	.140	.183	.107
		48	42	.093						
6	.138	32	36	.106	7/16	9/32	19	.166	.226	.132
		40	33	.113						
8	.184	32	29	.136	1/2	11/32	10	.193	.270	.156
		36	29	.136						
10	.190	24	25	.149	19/32	3/8	7/32	.218	.313	.180
		32	21	.159						
1/4	.250	20	7	.201	3/4	1/2	9/32	.281	.375	.216
		28	3	.213						
5/16	.312	18	F	.257	15/16	5/8	11/32	.343		
		24	I	.272						
3/8	.375	16	5/16	.312	1 1/8	3/4	13/32	.406		
		24	O	.332						
1/2	.500	13		.422	1 1/2	1	9/16	.562		
		20		.453						
5/8	.625	11		.531	1 7/8	1 1/4	11/16	.687		
		18		.578						
3/4	.750	10		.656	2 1/4	1 1/2	13/16	.812		

	Flat Head		Soc Head		Cap Screw	Nut		Washer	
	D	H	D	H	F	F	H	D	H
						0.156	0.047	0.156	0.020
								.078	.025
						.187	.063	1/4	1/32
						.187	.063	1/4	1/32
	0.225	0.048	0.183	0.112	0.079	.250	.093	5/16	1/32
	.279	.060	.226	.138	.094	.312	.109	3/8	1/32
	.332	.072	.270	.164	.127	.344	.125	3/8	1/32
	.385	.083	.312	.190	.158	.375	.125	7/16	.042
	.507	.110	.375	.250	.189	.438	.125	1/2	1/16
	.635	.138	.437	.312	.220	.500	.156	3/4	1/16
	.762	.165	.562	.375	.315	.562	.188	7/8	1/16
			.750	.500	.378	.750	.250	1 1/8	3/32
			.875	.625	.503	1.000	.312	1 1/4	3/32
			1.000	.750	.565			1 1/2	3/32

American Standard (Morse) Tapers.

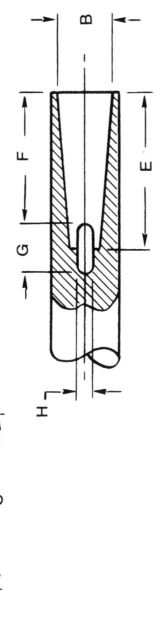

Taper Number	Taper Per Inch	Small End Diameter (A)	Large End Diameter (B)	Whole Length (C)	Depth (D)	Depth of Hole (E)	Top of Tang Slot (F)	Length of Tang Slot (G)	Width of Tang Slot (H)	Tangs Thickness (J)	Tang Length (K)
0	.052000	.252	.356	2.34	2.218	2.31	1.94	.56	.166	.156	.25
1	.049882	.369	.475	2.56	2.437	2.16	2.06	.75	.213	.203	.37
2	.049951	.572	.700	3.12	2.937	2.61	2.50	.87	.260	.250	.44
3	.050196	.778	.938	3.87	3.687	3.25	3.06	1.19	.322	.312	.56
4	.051938	1.020	1.231	4.87	4.675	4.12	3.87	1.25	.479	.469	.62
5	.052626	1.475	1.748	6.12	5.875	5.25	4.94	1.50	.635	.625	.75
6	.052137	2.116	2.494	8.56	8.250	7.33	7.00	1.75	.760	.750	1.12
7	.052000	2.750	3.270	11.62	11.250	10.08	9.50	2.87	1.135	1.125	1.37

268

Standard Fits.

Nominal Size Range Inches	Force Fit Clearance	Force Fit Hole	Force Fit Shaft	Location Fit Clearance	Location Fit Hole	Location Fit Shaft	Sliding Fit Clearance	Sliding Fit Hole	Sliding Fit Shaft	Close Running Fit Clearance	Close Running Fit Hole	Close Running Fit Shaft	Medium Running Fit Clearance	Medium Running Fit Hole	Medium Running Fit Shaft	Free Running Fit Clearance	Free Running Fit Hole	Free Running Fit Shaft
0 - 0.12	-0.05	+0.25	+0.5	0.3	+1.0	-0.3	0.1	+0.25	-0.1	0.3	+0.6	-0.3	0.6	+1.0	-0.6	1.0	+1.0	-1.0
	-0.5	0	+0.3	1.9	0.	-0.9	0.55	0	-0.3	1.3	0	-0.7	2.2	0	-1.2	2.6	0	-1.6
0.12 - 0.24	-0.1	+0.3	+0.6	0.4	+1.2	-0.4	0.15	+0.3	-0.15	0.4	+0.7	-0.4	0.8	+1.2	-0.8	1.2	+1.2	-1.2
	-0.6	0	+0.4	2.3	0	-1.1	0.65	0	-0.35	1.6	0	-0.9	2.7	0	-1.5	3.1	0	-1.9
0.24 - 0.40	-0.1	+0.4	+0.75	0.5	+1.4	-0.5	0.2	+0.4	-0.2	0.5	+0.9	-0.5	1.0	+1.4	-1.0	1.6	+1.4	-1.6
	-0.75	0	+0.5	2.8	0	-1.4	0.85	0	-0.45	2.0	0	-1.1	3.3	0	-1.9	3.9	0	-2.5
0.40 - 0.56	-0.1	+0.4	+0.8	0.6	+1.6	-0.6	0.25	+0.4	-0.25	0.6	+1.0	-0.6	1.2	+1.6	-1.2	2.0	+1.6	-2.0
	-0.8	0	+0.5	3.2	0	-1.6	0.95	0	-0.55	2.3	0	-1.3	3.8	0	-2.2	4.6	0	-3.0
0.56 - 0.71	-0.2	+0.4	+0.9	0.6	+1.6	-0.6	0.25	+0.4	-0.25	0.6	+1.0	-0.6	1.2	+1.6	-1.2	2.0	+1.6	-2.0
	-0.9	0	+0.6	3.2	0	-1.6	0.95	0	-0.55	2.3	0	-1.3	3.8	0	-2.2	4.6	0	-3.0
0.71 - 0.95	-0.2	+0.5	+1.1	0.8	+2.0	-0.8	0.3	+0.5	-0.3	0.8	+1.2	-0.8	1.6	+2.0	-1.6	2.5	+2.0	-2.5
	-1.1	0	+0.7	4.0	0	-2.0	1.2	0	-0.7	2.8	0	-1.6	4.8	0	-2.8	5.7	0	-3.7
0.95 - 1.19	-0.3	+0.5	+1.2	0.8	+2.0	-0.8	0.3	+0.5	-0.3	0.8	+1.2	-0.8	1.6	+2.0	-1.6	2.5	+2.0	-2.5
	-1.2	0	+0.8	4.0	0	-2.0	1.2	0	-0.7	2.8	0	-1.6	4.8	0	-2.8	5.7	0	-3.7
1.19 - 1.58	-0.3	+0.6	+1.3	1.0	+2.5	-1.0	0.4	+0.6	-0.4	1.0	+1.6	-1.0	2.0	+2.5	-2.0	3.0	+2.5	-3.0
	-1.3	0	+0.9	5.1	0	-2.6	1.4	0	-0.8	3.6	0	-2.0	6.1	0	-3.6	7.1	0	-4.6
1.58 - 1.97	-0.4	+0.6	+1.4	1.0	+2.5	-1.0	0.4	+0.6	-0.4	1.0	+1.6	-1.0	2.0	+2.5	-2.0	3.0	+2.5	-3.0
	-1.4	0	+1.0	5.1	0	-2.6	1.4	0	-0.8	3.6	0	-2.0	6.1	0	-3.6	7.1	0	-4.6
1.97 - 2.56	-0.6	+0.7	+1.8	1.2	+3.0	-1.0	0.4	+0.7	-0.4	1.2	+1.8	-1.2	2.5	+3.0	-2.5	4.0	+3.0	-4.0
	-1.8	0	+1.3	6.0	0	-3.0	1.6	0	-0.9	4.2	0	-2.4	7.3	0	-4.3	8.8	0	-5.8
2.56 - 3.15	-0.6	+0.7	+1.8	1.2	+3.0	-1.0	0.4	+0.7	-0.4	1.2	+1.8	-1.2	2.5	+3.0	-2.5	4.0	+3.0	-4.0
	-1.8	0	+1.3	6.0	0	-3.0	1.6	0	-0.9	4.2	0	-2.4	7.3	0	-4.3	8.8	0	-5.8

When fitting a cylindrical part into a cylindrical hole, the desired amount of clearance between the part and the hole will depend on how the part is to be used. An interference fit or force fit is desired for a permanent assembly. A slip fit is a snug fit that will permit easy disassembly. Running fits will permit freedom to rotate with allowances for part expansion resulting from frictional heat buildup. This table can be used to determine the proper tolerances for both the hole and the cylindrical part to give you the type of fit you want.

The dimensions shown in the table are thousandths of an inch (1.6 in the table equals 0.0016 inches). To use the table, select the proper size range in the first column at the left and find where that row intersects with the required "type of fit" column. The tolerances are then added or subtracted from the desired nominal diameter as indicated by the + and − signs. As an example, a force fit for a 3/16 (0.1875) nominal diameter pin would be a 0.1879 + .0002, − .0000 diameter pin in a 0.1875 +.0003, − .0000 diameter hole. A free running 3/8 (0.375) shaft and hole would be 0.3734 + .0000, − .0009 diameter shaft in a 0.3750 +.0014, − .0000 diameter hole.

The following formulas can be used to calculate the weight of materials or parts. All dimensions are in inches, W in the formula is material weight per cubic inch and is taken from the table below. All weights are in pounds.

Round bar stock

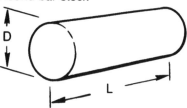

weight = 0.7854 × D² × L × W

Square bar stock

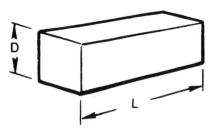

weight = D² × L × W

Hex bar stock

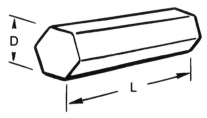

weight = 0.8662 × D² × L × W

Flat stock

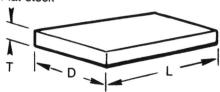

weight = T × D × L × W

Tubing

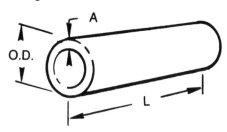

weight = 3.1413 × (O.D. − A) = L × W

Octagonal bar stock

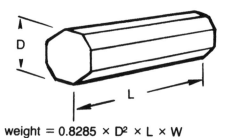

weight = 0.8285 × D² × L × W

WEIGHT IN Lbs. PER Cu. In.

Material	W	Material	W
1100 Alum.	0.098	Cast Iron	0.258
2011 Alum.	0.102	Carbon Steel	0.283
2014 Alum.	0.101	300 Stainless	0.286
2024 Alum.	0.100	400 Stainless	0.283
6061 Alum.	0.098	Lead	0.410
Brass	0.307	Silver	0.379
Copper	0.324	Gold	0.698

Circle

Area $= A = 3.1416 \times r^2$
Circumference $= C = 3.1416 \times d$
Diameter $= d = 2r$
Radius $= r = d/2$

Square

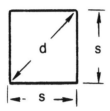

Area $= A = s^2 = \frac{1}{2}d^2$

Rectangle

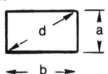

Area $= A = a \times b = a\,d^2 - a^2 = b\,d^2 - b^2$

Parallelogram

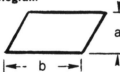

Area $= A = a \times b$ Note: dimension a is measured at right angles to line b.

Right triangles

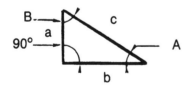

Area $= A = \dfrac{a \times b}{2}$

$a = c^2 - b^2 = c \times \text{Sin } A = b \times \text{Tan } A$

$b = c^2 - a^2 = c \times \text{Cos } A = a\,/\,\text{Tan } A$

$c = a^2 + b^2$

Sin $A = a/c$

Cos $A = b/c$

Tan $A = a/b$

Angle B $= 90° - A$

Angle A $= 90° - B$

The table is used to convert trig functions (Sin, Cos, and Tan) to angles and angles to trig functions.

Strength and Machinability of Common Materials.

The following chart gives the tensile strength and machinability ratings for some common materials. The values shown are approximate and should be used for comparison only. These values are very dependent upon processing differences and heat-treat conditions.

Machinability ratings are in respect to standards, all ferrous alloys are compared to steel alloy 1112, aluminum alloys are compared to aluminum alloy 2011, and copper alloys are compared to brass alloy 360.

Material	Alloy	Tensile Strength PSI	Machinability
Iron		40,000	
Steel	1025	60,000-103,000	50%
	1045	80,000-182,000	55%
	1095	90,000-213,000	45%
	1112	60,000-100,000	100%
	1212	57,000-80,000	100%
	4130	81,000-179,000	58%
Aluminum	2011-T8	57,000-62,000	100%
Alloys	2024-0	24,000-28,000	50%
	2024-T4	60,000-68,000	90%
	6061-0	18,000-19,000	50%
	6061-T6	45,000-46,000	75%
Stainless	301	100,000-185,000	45%
Steel	302 & 304	75,000-110,000	45%
	303su	90,000-125,000	60%
	410	95,000	54%
	416	70,000	97%
	440	110,000	39%
Copper Alloys			
pure copper	101	32,000-48,000	54%
brass	360	47,000-58,000	100%
	3532	45,000-73,000	90%
	3711	60,000	90%
	464	56,000-63,000	30%
bronze	4611	56,000-63,000	30%
	6421	88,000	60%

Index

Made in the USA
San Bernardino, CA
13 September 2017